PROJECTION DESIGN

THE BASICS

Projection Design: The Basics explores the concepts of visual elements in live entertainment. It provides a conversational view of the fundamentals of projection design, from production meetings and the elements of visual design to the equipment necessary to make it all happen.

This text examines the themes and theories universal to a wide range of topics, to provide a foundation for anyone interested in using video for their live production or for those who are looking where to start as a designer. Topics covered include:

- Methods of extracting visuals from a script and communicating them to production staff.
- Basics of visual design.
- Understanding human perception and how this influences design.
- How to choose the right equipment to build a system.

With a detailed glossary, basic formulas, and comprehensive explanation from start to finish, *Projection Design: The Basics* is an ideal primer for Projection Design courses, and will be of interest to anyone entering the field of projection and media design for the first time.

Davin E. Gaddy has been working behind the scenes of productions since 1988, from professional sporting events to Broadway tours and everything in between. He shares his experience with students and professionals alike through mentorships, lectures, and workshops. He is the author of *Media Design and Technology for Live Entertainment*.

The Basics Series

The Basics is a highly successful series of accessible guidebooks which provide an overview of the fundamental principles of a subject area in a jargon-free and undaunting format.

Intended for students approaching a subject for the first time, the books both introduce the essentials of a subject and provide an ideal springboard for further study. With over 50 titles spanning subjects from artificial intelligence (AI) to women's studies, *The Basics* are an ideal starting point for students seeking to understand a subject area.

Each text comes with recommendations for further study and gradually introduces the complexities and nuances within a subject.

STATISTICAL ANALYSIS
CHRISTER THRANE

ANTHROPOLOGY OF REPRODUCTION
SALLIE HAN AND CECÍLIA TOMORI

SCIENCE COMMUNICATION
MASSIMIANO BUCCHI AND BRIAN TRENCH

PROJECTION DESIGN
DAVIN E. GADDY

For more information about this series, please visit: www.routledge.com/The-Basics/book-series/B

PROJECTION DESIGN

THE BASICS

Davin E. Gaddy

NEW YORK AND LONDON

Designed cover image: WICKED focus grid, 2003, Elaine J. McCarthy and Jacob Pinholster

First published 2025
by Routledge
605 Third Avenue, New York, NY 10158

and by Routledge
4 Park Square, Milton Park, Abingdon, Oxon, OX14 4RN

Routledge is an imprint of the Taylor & Francis Group, an informa business

© 2025 Davin E. Gaddy

The right of Davin E. Gaddy to be identified as author of this work has been asserted in accordance with sections 77 and 78 of the Copyright, Designs and Patents Act 1988.

All rights reserved. No part of this book may be reprinted or reproduced or utilised in any form or by any electronic, mechanical, or other means, now known or hereafter invented, including photocopying and recording, or in any information storage or retrieval system, without permission in writing from the publishers.

Trademark notice: Product or corporate names may be trademarks or registered trademarks, and are used only for identification and explanation without intent to infringe.

ISBN: 978-0-367-61874-2 (hbk)
ISBN: 978-0-367-61861-2 (pbk)
ISBN: 978-1-003-10701-9 (ebk)

DOI: 10.4324/9781003107019

Typeset in Bembo
by Newgen Publishing UK

CONTENTS

List of illustrations — vi
About the cover image — viii
Preface — xi
Acknowledgments — xiii

1 What is projection design? — 1
2 Expectation and perception: 2D in a 3D world — 19
3 Design elements and workflow — 41
4 Basic concepts — 64
5 Content — 88
6 Equipment — 120
7 Show control — 150
8 Building a system — 167

Glossary — 198
Index — 211

ILLUSTRATIONS

FIGURES

0.1	*Wicked* focus grid, 2003	viii
1.1	Hand shadow puppets as seen in the 1861 volume of *Le Magasin Pittoresque*	12
1.2	An example of the magic lantern as used in a phantasmagoria	13
1.3	Patent for zoetrope	15
2.1	The electromagnetic spectrum	21
2.2	Methods of reflection and refraction	25
2.3	Anatomy of the eye	28
4.1	*The Ponte Salario* by Hubert Robert	65
4.2	*Wreath of Laurel, Palm, and Juniper* by Leonardo da Vinci	67
4.3	A simple optical illusion	71
4.4	A more complex illusion, illustrating the Bartleson-Breneman effect	72
4.5	*The Larder* by Anton Maria Vassallo	74
4.6	*The Marketplace in Bergen op Zoom* attributed to Abel Grimmer	76
4.7	*Wuxtry!* by Albert Abramovitz	80
4.8	*Moonlight* by Claude-Joseph Vernet	86

5.1	*A Sunday on La Grande Jatte* by Georges Seurat	101
6.1	Block representation of light dispersion	128
8.1	Three potential projector positions	172
8.2	Optic design allowing for off-axis projection	181
8.3	Example of digital processing to correct keystone	182
8.4	The process of edge blending	186
8.5	Hot spots with edge- blended rear projection	188
8.6	Illustrating mismatched aspect ratios	193

TABLE

5.1	Standard display resolutions and their associated information	110

ABOUT THE COVER IMAGE

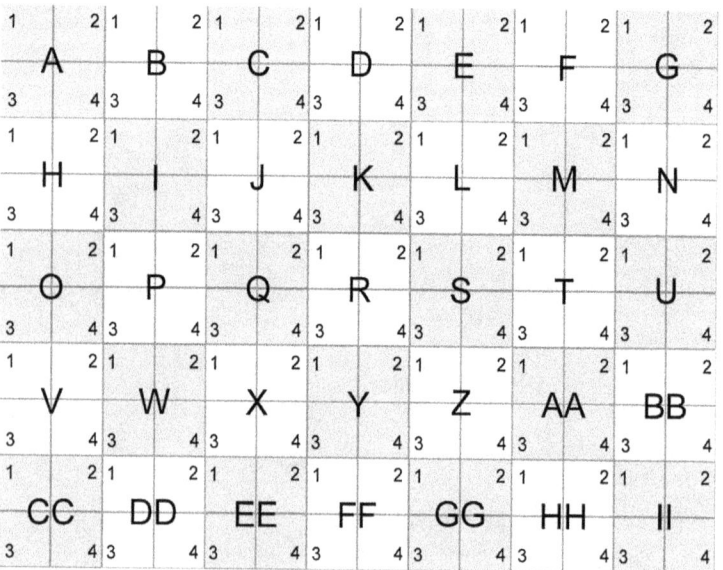

Figure 0.1 *Wicked* focus grid, 2003.
Source: Elaine J. McCarthy and Jacob Pinholster.

ABOUT THE COVER IMAGE

The grid grounds you. It is a baseline reference – a known entity – a map. This grid maps a certain time and place in the pixel playground – the opening of *Wicked* on Broadway; typical projector resolutions of 1920 pixels x 1080 pixels, 72 pixels/inch; the use of the beta version of the Catalyst media servers and the use of moving mirror heads in projections. The green grid has gone on to become a metaphor for the expansion of the use of projection design in live entertainment over the past 20 years. Jake Pinholster, who built it for me as my assistant designer, made it available to his students and online through *Live Design* magazine; and I suspect it has gone out on a few computers from shops that forgot to delete it. His students used it and passed it on to colleagues and their own students. The green grid gets around – it has been spotted from Detroit to Dubai. If it comes into your life, feel free to send me a photo of it in use at: wicked.green.grid@gmail.com.

Elaine J. McCarthy

PREFACE

The 21st century has led to amazing developments in the world of entertainment. Live entertainment certainly has competition when compared to the gamut of other options, from cinema to video games to the volumes of social media platforms. One technology that tends to make the greatest impact in keeping audiences coming to seek out live options is that of video. LED displays are becoming a common sight in everyday life, often in the form of digital signage. Creating something special is still a possibility.

Even though creatives have been using light to create images for thousands of years, technology to specially create pictures with light has been a focus in the past few centuries. Recently, this technology has become much more accessible to the masses. In the past few years, it has exploded, with more and more creators and audiences looking for it. The question comes to: "How do we do it?"

In this book, we look to share the foundation of projection design. With the fundamentals firmly cemented, the budding projection designer will be able to grow, no matter how the technology changes. Due to this, generalized descriptions of equipment and software will be used as opposed to naming specifics which may not be available in the near future. It is more important to understand the concepts behind how the design is realized than what the most popular hardware or software may be at the time of publication. In order for a successful design, there must be an understanding of not only the processes going into the project, but also how each component works and how to tie them together.

The strategy of this style of introduction to the topic of projection design will open possibilities for years to come. Instead of being an instruction manual with step-by-step procedures, which can rapidly become dated or stifle creativity, this approach is a guideline to success. It is the author's hope that the years of successfully mentoring projection students translates to the page. The principles set forth in the following chapters give students of projection the tools necessary to learn how to grow with this field, especially as artificial intelligence continues to creep into digital media.

ACKNOWLEDGMENTS

When I was young, my life goal was that I would one day discover Atlantis. I wanted to dig through the lore and mythology to find the truth behind the legend. But life does not always lead us in the direction we think it will. Even when I started in live entertainment, I thought that I would be a rigger or lighting technician, having no idea that projection was soon going to envelop my life.

For my career, I want to thank those who have taken a chance on someone who avoids keeping records of their accomplishments. I know that I proved myself worthy of those leaps of faith as continued opportunities were offered my way.

Within the first couple of months of working as the lead projectionist of KÀ, by Cirque du Soleil, I was asked to share my experiences with a group of college students. That was just the beginning. It is possibly due to me saying yes at that time that we find ourselves here today. I want to thank the educators who have invited me to share my knowledge with the upcoming generations of designers and technicians. I particularly want to thank the students for their yearning to explore the possibilities that photonic alchemy can provide. Each question not only helped them grow, but opened my eyes to more ways of seeing this amazing field of production.

Specifically, I want to thank Wendall K. Harrington and Jake Pinholster who supported my scholastic journey. After being asked to teach workshops, Ms. Harrington kindly provided me with guidance and in particular encouraged my first foray into authorship, which resulted in *Media Design and Technology for Live Entertainment*. Mr. Pinholster was essential in helping me to organize my experience into something useful for lectures and workshops.

I want to thank the Digital Media Commission of USITT for allowing my continued participation in the program. The honor of leading the charge of creating the ESET exam for projection was another opportunity for my own personal growth while helping the community as a whole.

Finally, I want to thank my personal teachers who encouraged the concept of learning how to learn. It is because of this skill that I was able to transition my career and maybe, just maybe, I have learned to bring the magic and mystery of Atlantis to the modern world through the magic of projection.

WHAT IS PROJECTION DESIGN?

42. This simple answer is likely confusing. While this is lifted from *The Hitchhiker's Guide to the Galaxy* by Douglas Adams for a completely separate question (the ultimate question of life, the universe, and everything), this answer can be just as appropriate for projection design (more on this at the end of the chapter). In his book series, Adams identifies the issue of properly identifying the question, which was more complicated than might be assumed. Also, it presents guidance in that sometimes we need to know what we want to achieve and then work backwards. As this text is not a book on philosophy, we will avoid some specifics, knowing that this is an evolving field and that the dynamics will fluctuate over time. It will be emphasized that knowing what the appropriate question is during the design process is critically important.

If provided by the audience, the answer to what projection design is will most often simply be "magic." Indeed, the optical principles that are fundamental to modern projection have been studied and applied for hundreds of years. When properly applied, the audience may not be able to perceive where the physical ends and the projection begins, as was the case in the Broadway production of *Beetlejuice*. Thus, the projection designer may want to understand the physical properties of projected light as well as the psychology of how imagery impacts the audience, in addition to

DOI: 10.4324/9781003107019-1

understanding the way humans perceive light. These elements, when properly applied, will transport the audience to the next realm. And yes, the producer and director may think that the designer has conjured up something spectacular as they themselves may not understand what it takes to pull off the design.

In North America, USA 829, the designers' branch of the International Alliance of Theatrical Stage Employees, Moving Picture Technicians, Artists and Allied Crafts of the United States, Its Territories and Canada (IATSE) defines a projection designer as one who creates "ephemeral imagery for live performance delivered via a wide variety of technologies including projectors, LED surfaces, monitors, and other digital and analog methods." As they clearly state, projection design can include technologies beyond the video projector. Due to this, in productions outside of the union's influence, the projection designer may be identified by a number of different titles, most often as Media Designer. The Digital Media Commission of the United States Institute of Theatrical Technology (USITT) is a body that is looking to additionally develop best practices to advance the field. This is to say that this evolving field will continue to fluctuate for years to come while the basic principles will remain a steady foundation.

Building from the IATSE definition, let us look at the common fundamentals that are involved, regardless of what title is given to the design. A projection design is ephemeral. It is extremely temporary. As soon as the light source is gone, only the memory remains. This will be important to remember later on. Though it is not limited to video projectors, specifically not even restricted to digital devices, the temporary nature is extremely important. The designer has the use of many different tools, including variations among similar tools, which can impact how the design is perceived. The fundamentals of what is to be displayed will remain the same, but how the designer goes about displaying it will make a difference.

Beyond the "how" a design is achieved, is the "for what" it shall be used. Projection design is a tool to help advance the story as the director sees fit. There are a number of choices that the designer may make to fulfill that goal. The projection designer is not constrained by a projection screen, as would be the case in cinema. If the design is to be projected, then the only constraint is what it is to be projected on (hint, it does not need to be a flat rectangle). This is part of the freedom that this media provides; it can work on so many levels, offering a new voice to storytelling. It allows a greater immersion into the world of the production that was scarcely imagined a few decades ago.

Is projection design a form of scenic design? In a manner of speaking, yes. One reason that many directors ask for projection design is to replace scenic backgrounds. Some vendors who provide scenic drops have started to digitize many of their designs, in order for them to be projected and animated, but it can upstage the performance if not done correctly. Projection can be used to add texture to fairly untextured scenic elements. How the surface of the scenery is treated will have an immense impact on the quality of the image. But projection design is so much more than just a part of scenic design. With just these couple of examples, it is evident that the projection designer and scenic designer will work closely together.

Is projection design simply another lighting design? In a manner of speaking, yes. Projection by its very nature is creating images with light. Some imaging devices are truly specialized lighting instruments. Consider that a slide projector is very similar to an ellipsoidal reflector spotlight (commonly called a Leko) or profile fixture in how they create an image, putting an image medium between the light source and the lens. It is also a possibility that the projectors may be the best lighting units for specials, creating lighting effects not possible with traditional lighting units. But projection design is so much more than just being a part of a lighting design. This is why the projection designer and lighting designer will typically work closely together, especially as they will be in direct competition for providing light.

Can the projection design become a performer? In a manner of speaking, yes. When choosing the media and how it will be presented, the choice does not simply need to be something in the background. There are several techniques where this medium can become interactive or have a life of its own. These are often some of the "wow" moments that happen in talent game shows (such as Front Pictures' performance on *America's Got Talent*). As gaming engines (such as Unreal Engine or Unity) are used to enhance projection design, more digital puppets and avatars will be present in future productions. However, a projection design is often used to enhance a production subtly and is often almost imperceptible.

The use of light to produce images is not new to entertainment. What is new is the accessibility and relative ease of doing so. To answer the question of what is projection design will require answering a lot more questions; and some of those questions are being redefined every day. The easy answer is that it is the part of a theatrical or entertainment production that most often utilizes video projectors to place an image (still or moving) on a part of the set (or performer) to support a production. But it is a lot more, and over the course of this book there is no one set universal solution. So let us take a look at what might be needed to create a design like this, regardless of the role you might play.

WHAT DOES IT TAKE?

There is a variety of equipment to present the design, from the source to the display(s). There is no one set formula for what will be needed. There is, however, a complex set of variables that can proffer what may be anything from a mediocre to a splendid design. With projection design there is the content and the method of creating that content. The content is the video or image files that will be presented via some display device. Also, there is the system of devices to manipulate the content and allow it to be seen by the audience as well as the space in which it will be presented.

As will be discussed in later chapters, there are three main components of a projection system technology: the source, the method of distribution, and the display. The complexity of the system can be simple or advanced, based on a number of factors. Sometimes the availability of equipment on hand will determine how the design will be built, while other times may allow for custom systems. Much will depend on the type of production that is being created. Will this be a traditional play, a musical, a concert, or some other type of event? Each type of production will have different expectations and needs.

The projection designer, like all designers, will need to meet with the director to determine the needs of the production. This could be designing the content to fit an installed system or, in some cases, "absolutely everything." Both aspects will be approached as the principles are the same; they just vary in scope.

TOOLS

For many productions, the content will be stored and played back on a computer or network of computers. The computer will have some form of program that will manipulate the content, even if it simply allows the operator to easily play the video in time with a cue. These computers and their programs are commonly known as media servers. Less commonly, other media players are used; these include optical devices such as DVD or Blu-ray players, but more likely a solid-state media player. Gone are the days of magnetic tape players and most other analog media.

The most obvious part is the display, that which allows the audience to see the content. As has been suggested by the title of the book, some sort of video projector is the most common display. Not all projectors are the same, as they use different light sources and ways of creating the image, but there are common specifications that will need to be understood for a quality presentation. As production budgets increase, the range of displays may include some sort of video wall, most often a light-emitting diode (LED). Also, depending on the production, the display may need some practical effect, like an old cathode-ray tube (CRT) television or computer monitor, which may create new challenges to get signal to them.

How the source connects to the display requires some form of distribution. This may be a simple cable, allowing for the source and the display to communicate, or it could be combined with a variety of intermediary devices to get the signal to the display(s). While a single source and a single projector may be all that is required for a production, the distance between the two may require something more than a simple cable can handle due to technical restrictions (more on this in Chapter 6). Some designs will have multiple displays that either have the same content or completely separate content; how that gets from the source to the display can happen in some very complex configurations.

COOPERATION

The designer will need collaboration with other departments. The placement of the displays will impact rigging, carpentry, lighting, and audio. Projectors are often in need of being rigged off the ground so as to reduce shadows and maximize visibility of the image. Unlike theatrical lighting, the precision necessary does not always allow them to be supported on standard theatrical rigging, nor is there a standard method for safely rigging them. In order to make sure that they do not get in the way of lighting or moving scenic units, do not have obstructions in the path of the beam, and do not have anything to impact the stability of the unit, considerable planning will need to take place. When the projection is on the set, how the set is painted and textured will impact how the image will look. If the projector is on some movable position (a lighting batten or moving scenic piece), precision and repeatability are critical for the image placement to be replicated. In addition, projectors require cooling, and the noise of the fans will need to be taken into consideration as to how that can impact the sound design.

WHY DO WE DO IT?

Why does the director choose to use projection? Projection can magically transport the audience to a new world, especially when combined with automation and audio effects through show control: the centralized management and coordination of various technical components of live entertainment. If it is a theatrical production, some sort of video may be required by the script, even if just as images played on a prop television. There is also the possibility that the director hopes to replace some scenic elements such as a backdrop. It is understandable why there may be a desire to bring a projection design into a production, even if the desire may be purely a function of potential cost savings. Hopefully, projection is not being used just to up the "production value" of a show by highlighting new technology.

Video can have very practical reasons as to why it is required while other times it can be purely aesthetic. It can provide information or scenery, a special effect, or bring the audience closer to the action. How it is being utilized should guide the designer in not only creating the content, but also how that content will be displayed. How the design should look will help the designer choose the type of display, how many displays will be required, how the content will be handled (source and distribution). This is one of the ways in which the designer will work backwards, having a feeling for what the end goal should be and building to reach that goal.

There are essentially six functions that video can serve in a production (discussed in detail in Chapter 5). Some designs may use one, while others use all of them. Some designs may be able to function with a single projector covering a wide gamut of uses, while others may require multiple displays, some only being used for a single purpose. The functions of projection design are not limitations, but classifications which may provide understanding of how they may be achieved.

WHO IS INVOLVED?

It may seem obvious that there are certain people involved in how a design is achieved, such as the director and the projection designer, but the team and whom they impact may be less obvious. In order to have a successful design, the cooperation of departments is essential, but as projection design is not always understood, flexibility should be kept in mind as well, especially when budgeting time during installation and final rehearsals. This list of parties involved will be parsed towards large productions to best include all possible job duties, which are often combined into fewer staff on smaller shows.

THE DIRECTOR

The director is the person who artistically drives the production. This person not only tells the performers what to do, but also how the overall show will look and sound. The downside is that many directors do not understand how to make their vision happen on the technical side of things or if it can happen at all. They will rely on the expertise of the team that they have assembled to fulfill their decisions. Due to this, the projection designer needs to have excellent communication skills in order to inform the director of what is possible and how it will be accomplished (or alternative solutions, especially when budget is concerned).

As the director is like the CEO of a corporation, a primary duty is to guide the coordination of departments in the production. The director will build their team to design the show. The first person to be hired will be the general stage manager. This person is responsible for the show after it opens and must be present to understand the wishes of the director. Next, very early in the process, the designers will be brought on board. Hopefully, especially as projection design becomes a better-understood part of production, the projection designer will be brought in at the same time as the other designers. Unfortunately, this is not always the case in semi-professional and novice productions.

THE SCENIC DESIGNER

A production on a bare stage is often not the first choice for most directors; a scenic designer is usually one of the first of the design team to be hired. This person is responsible for the general stage elements that define the space into representations of different places. As already mentioned, the traditional scenic element of the backdrop is sometimes replaced with a digital version. If this is all that will be done with projection, especially if the decision is to use a service that specializes in digital backgrounds, then the director may give the scenic designer the responsibilities of projection. There are many reasons a projection designer may not be hired, this being a specific instance.

Besides projecting on backdrops, the scenic designer is providing the surface for a variety of projection. The material choices of those scenic elements and how they are finished can have an immense impact on how the light will behave. These materials will include soft goods (drops, scrims, and curtains) to project on or to absorb light. This may require considerable testing during early parts of the production to make sure that there are few surprises as the production nears opening. It is recommended that the specific projectors to be used in the production be used during any tests as the light source for each projector is slightly different and may change the outcome. This is easiest when the production owns the projectors as the cost of rentals can be quite high and budget may make this prohibitive.

THE LIGHTING DESIGNER

As a rule of thumb, productions held in a theater have no natural light; a lighting designer will create how a production looks through the use of many different types of lighting instruments placed throughout the venue to illuminate the performers and the scenic elements. While shining a light directly on a backdrop will be required for it to be seen, any unwanted light on a projection surface will be detrimental to its overall look.

It is possible that simple scenic projections may be handled by the scenic designer as stated above, but if a projection designer is not hired, these duties are more likely to be given to the lighting designer than the scenic designer. Since projectors and LED walls all emit light, it may be considered common sense that the lighting designer take charge of the projection, especially when it is a pre-packaged service of content for a known production. At that point, the lighting designer is likely only working on how to control the content and keep light off the projection surface, as the "design" has been done by the provider.

THE VIDEO TEAM

Rarely will a projection designer work completely solo. The larger the production, the greater number of support staff will likely be needed. The size and scope of the video team will depend on many factors, including the amount of content which needs to be created (animator or video producer), the complexity of the media server (programmer), the complexity of distribution to the displays (video engineer), and the skills required to master the manipulation of photons (the projectionist). There is no set formula for how many of each discipline the designer may need to employ. It will come down to how the designer needs to delegate the tasks; including whether or not the responsibilities are passed on to another department (such as the lighting operator controlling the media servers).

The content creation team is going to be brought in by the designer as early as possible. The size of this team will vary depending on how much content needs to be created and edited, and ultimately on the budget. If content requires filming the performers in their costumes, work with the costume designer will be crucial to get the wardrobe, wigs, and makeup set as early as possible for continuity. The designer may also contract additional firms which specialize in content creation in order to get some of the bulk work done while keeping the in-house team free for specialized projects which may require more refinement.

The designer will bring to the team specialists in displays and signal distribution. These individuals will be able to choose the equipment necessary to make the designer's vision come to life. This will include understanding which projectors to pick that will be bright enough and the appropriate lenses to choose for image size; and possibly other features – for example, edge blending for using multiple projectors to create a larger image. They will also be able to design the system to get the content to the appropriate projector, especially if multiple displays need to be used for a single image.

The person responsible for operating the production may or may not be the individual responsible for programming the media servers or setting up the rest of the equipment. The operator is generally the person responsible for maintaining the designer's vision after the production has opened. The complexity of programming the media servers might be beyond the scope of the operator. As with the programmer for the lighting designer, the video programmer will be a specialist in a particular system and be able to interpret the desires of the designer and manipulate the control system to make it happen. They will be able to not only enhance the look of content more than just playing back a clip, but they can also prepare the video to operate with a show control system, such as timecode (a sequential numerical coded signal used for synchronization).

WHEN HAS IT BEEN USED?

While video projection is a fairly new art form, projection has been used in some form since the dawn of communication. Shadow puppetry likely came from a time before the written word. This can be accomplished in two manners which harken to the two main forms of modern projection, front and rear. In the former, the light source is on the same side of the projection surface as the audience. The light is manipulated by blocking a portion that is to hit a nearby surface, generally with the hands, molding the absence of light into recognizable shapes. This requires a single point source of light to create defined shadows. Alternatively, the light source is behind a translucent surface which diffuses the light, such as a piece of cloth, and the means of blocking it is also behind that surface. While the former will often use the storyteller's hands to create shapes (ombromanie or shadowgraphy), the latter has been popular in Asian countries using silhouette figures with rods to manipulate their appendages.

Fast forward several millennia to the Age of Enlightenment, where in Europe another means of manipulating light for the entertainment (and psychological manipulation) of audiences was created. The phantasmagoria, a type of production using a rudimentary projector known as a magic lantern or "lanterna magica," was popularized in the 18th century. These were done for amusement in allowing the audience to experience the supernatural. Haunting images were often the theme of the shows. However, the magic lantern itself was devised many years prior. Early prototypes are seen in the drawings of Leonardo da Vinci and others, possibly even back to ancient Greece, but the inventor most accepted for its creation is Christiaan Huygens in the mid-17th century. Many optical effects were created using mechanical manipulation of slides in the magic lantern. Slides which moved along a horizontal axis could create effects of moving water (similar devices are still used to this day for this effect) or other moving

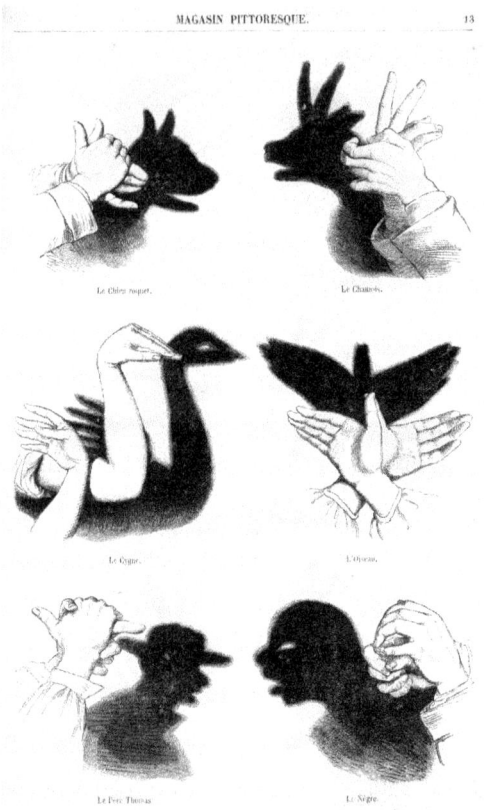

Figure 1.1 Hand shadow puppets as seen in the 1861 volume of *Le Magasin Pittoresque*.

Source: Image in public domain.

scenery. Levers, manipulating a second slide, could portray movement of body parts or reveal hidden elements. For a scenic element, such as a windmill, a pulley mechanism could rotate a slide. Other variations used up to three optical units, fading between one another for added effect.

Magic lanterns' popularity was finally superseded by the motion picture, though cinematic projection did not replace the magic lantern immediately. The 35mm slide projector, which is the modern variant and popularized in the 1950s, has even been used in fairly recent productions, such as the 54 projectors used in the 1992 production of The Who's *Tommy* (projection design by Wendall K. Harrington). In addition, large-format slide projectors were the answer when brightness was a requirement, until the end of the 20th century.

Motion, or the ability to help make the projected image come to life, has been the goal of inventors of projection. A variety of

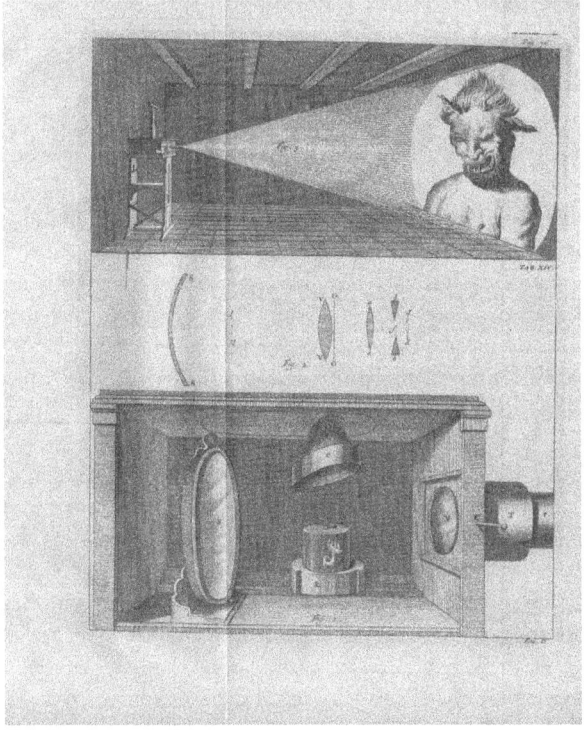

Figure 1.2 An example of the magic lantern as used in a phantasmagoria.
Source: Courtesy of the Max Planck Institute for Science; image in public domain.

novelty toys (based on scientific observations) had proven the concept that our eyes can be tricked into seeing motion from a series of still images. We see motion through the sequence of images by a process known as "persistence of vision," which is the optical illusion where a perceived image is retained after the light creating it has ceased to enter the eye. These toys were intended for the single observer, such as the phénakistiscope, which utilizes spinning disks with one having images in a looping sequence and the other contains a series of open panes, allowing the observer to see "motion." Possibly the more familiar variant of the phénakistiscope is the zoetrope, which is cylindrical. Variations on this using a source of light, known as the praxinoscope, produced some of the first motion pictures that could be projected. These all use a stroboscopic effect, allowing the viewer to momentarily see an image, then to have it blocked, then to see the next image. Having the view of the images blocked tricked the eye into seeing separate images as opposed to having them blur together. Even if the modern designer has never played with one of these toys, they have likely created or played with a flip book, which operates on the same principle as a stroboscopic device. The inventor of the praxinoscope, Charles-Émile Reynaud, adjusted the medium from a cylinder to a cloth strip with painted gelatin images spaced along it, creating the first movie. With the advent of flexible film media, the motion picture was able to replace the popularity of the magic lantern.

Though the cinématographe was not originally invented by the Lumière brothers, they made many significant improvements on it. While the screenings for the public were very short, less than a minute in length per feature, the films were no longer hand-painted representations that all previous projection had been, but now photographic imagery added life in a whole new perspective. Outside of the special showings of this new entertainment setting, the use in live production had not yet been realized. However, the cinema had additional developments in how the projection was presented, through various screen treatments to aid in viewing the projected image which have had a direct impact on modern screens and surface treatments.

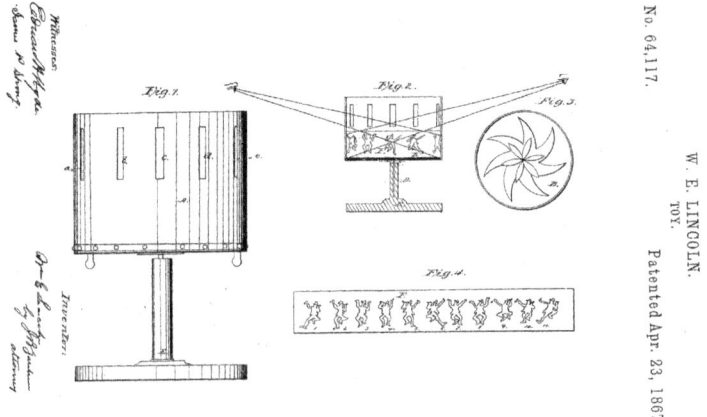

Figure 1.3 Patent for zoetrope.
Source: Image in public domain.

HOW IS IT ACCOMPLISHED?

As mentioned previously, there are three main components to the physical design – the source, the display, and the components of distributing content between those two. What is expected as the end result will depend on the individual. We may look at the projection on the castle at one of the Disney parks and think that we can come up with something similar on a much smaller budget. However, we should understand that our own castle (or other projection surface) will probably not look the same, as we will not be able to display as much light or as many pixels (a pixel is the smallest picture element created by a display); so we should not expect it to look the same. On the other hand, by learning how we as humans see, the designer can make some tweaks to the design to trick the observer into thinking the design is something other than initially perceived.

It is often said, "Content is king." After all, without content, all of our displays would just be showing video black (or snow if we have some old analog displays in use). Content comes in many forms, including physical media for some analog media such as film projectors. With media servers and other digital media players, the content will be required to have very specific ways that it is provided. These may be altered after they were originally created, but may suffer some loss of quality. So, we will discuss quite a bit about the creation of content, beyond just what it will look like.

Once the content is created, we want that content to be seen. We will look at the various types of equipment involved in displaying the content. This will run the gamut from discount projectors at big box stores to large venue projectors used to light up the castle. In addition, we will need to explore the challenges of getting the video signal from point A to point B and if there is anything that needs to happen along the way.

So many productions need video to coincide exactly with other show elements. As the audio design is often much more complex than a simple stereo mix, the audio is rarely embedded with the video signal. In some cases, having audio present can cause unexpected results in video transmission. So there needs to be some way to have the audio and video be in sync, potentially with other automated effects. This is where another element, show control, comes into play. Understanding how these elements work together is crucial. While there are entire books written about show control, this will just be an opportunity to understand what types of show control are available and how they play into the projection design.

Finally, we will look at what it takes to put it all together. There are so many different ways that we can shine a light on the subject, which is why the designer has the opportunity to ply the craft. However, to someone who does not have the technical savvy, knowing where to start can be daunting, to say the least.

COMMUNICATION AND VOCABULARY

In order to have successful communication, the designer and others involved need to be able to speak on the same terms. The basic concepts of video need to be understood by all involved, including the technology and how video is used. The director does not need to be an expert on all things video, but if the designer is unable to describe their requirements, then the design may fall short due to budgetary or other easily avoidable concerns.

Common language can often muddle communication. The simple word "standard" is used in everyday descriptions of how things are perceived to be as accepted practice. However, what one person thinks is accepted may not in actuality have a written and commonly accepted means of doing something. A standard, as defined by the International Organization for Standardization (ISO), is "a document that provides requirements, specifications, guidelines, or characteristics that can be used consistently to ensure that materials, products, processors, and services are fit for their purpose." In other words, a standard is a formula that describes the best way of doing something. It is more rooted in math and science, being more precise and defensible, than just a commonly held practice. The terms throughout this text, including the glossary at the end, will greatly enhance the ability of the designer and director to effectively understand one another.

Video displays are a regular part of the consumer market. The terminology and key points used by the sales associate at the local big box electronics store may have no bearing on what is required by a live production. Taking the time to understand that same terminology and how it applies to theatrical gear may give the designer an additional tool when trying to talk to other members of the creative team.

It is not just the consumer sales person who can make communication challenging. When the designer makes specifications on equipment based on properties required, they need to make sure that these specifications are accurate. Some of this has to do with the fact that not every manufacturer specifies their equipment the same way. Some may offer measurements based on published standards, while others do not. This is often the case with contrast ratio, the comparison of white to black illumination.

Getting back to the original answer to "What is Projection Design?" there is a little bit of a joke that Douglas Adams threw in there for us. The number 42 is the American Standard Code for Information Interchange (ASCII) designation for an asterisk. An asterisk is used in a search for everything of a certain file type, such as ★.doc to search for all document files in a given location. Given that the asterisk would search for everything, this would be a great ultimate answer. For projection design, it can also mean everything to questions like "What can I project on?"

> ### BOX 1.1 SUMMARY
>
> - Projection design is an evolving field which encompasses a variety of live entertainment fields.
> - Projection design can also be known as media design as it utilizes many different displays, not just video projectors.
> - Multiple disciplines of the production can have an impact on the design.
> - Projection designs may require a team to complete as they can become too complex to be actualized by a single individual.
> - There are three main components to a projection system: display, distribution, and source.
> - The history of projection design owes a lot to the showmen of the 18th and 19th centuries who created amazing visuals with what is now extremely rudimentary equipment.
> - Understanding vocabulary of the field is essential to effective communication.

EXPECTATION AND PERCEPTION
2D in a 3D world

Video, translated from Latin, is "I see."

For those who do not often work with video, some aspects that may seem familiar will actually be foreign. We will get into that in a bit. It used to be a common joke about the difficulty in connecting a video cassette recorder (VCR) to a television, which resulted in the clock never being set. In reality, the constant flashing of the time 12:00 really had little to do with the operation of the equipment. With modern consumer video equipment, connections are often limited to High-Definition Multimedia Interface (HDMI) and the equipment mostly makes all of the connections hassle-free, what is often referred to as "plug and play." Due to this, a lot of assumptions are made about how video works – assumptions which can lead to the disappointment of producers and directors alike, often due to the time it takes to install and make adjustments.

Connections and signal distribution will be discussed later, as will the variety of display technologies for the imagery to be seen. All of it can be challenges in the expectations of how a projection design might come to fruition. In order to fully understand what to use for the audience to see what is desired, the designer should understand a little bit about light (no matter how much we want to change it, light behaves in a predictable manner). Thus, it is necessary to have a basic understanding of how humans perceive light.

DOI: 10.4324/9781003107019-2

This knowledge will help to prepare in describing a successful design to those who do not comprehend the realizations and limitations. With more designs will come the ability to better predict how it will look, also giving the ability to communicate this to the rest of the artistic team. "Happy little accidents" will still happen, as there will almost always be a variable that wasn't considered (or changes were made) that give a little sparkle to the design that was unexpected.

PHYSICS OF LIGHT

Centuries of observation help us to know how light behaves. This allows us to plan on how we will utilize it. Lighting designers have had decades of experience in artistically crafting with photons and a variety of media between the light and its intended target (color and patterns). They understand that the color of the light source will impact the color media used and even how the source of light will affect how scenery and costumes will look.

What is light, after all? What we commonly call light is actually a very narrow part of the electromagnetic spectrum. This range is known as the visible spectrum due to what is perceptible to the human eye. The electromagnetic radiation that we can see falls between the wavelengths from about 380 to 750 nanometers and can vary by individual. The lower the frequency of wavelength, the color ranges into ultraviolet, while the higher frequency moves towards infrared.

A light source is comprised of a number of wavelengths in varying amplitudes, and when combined, forms white light. This white light will not always look the same from all sources and is thus described in color temperature. As these various sources of light are shone on a white surface, they will appear to have a tint towards colors of the red spectrum (warm) or towards the blue spectrum (cool). The correlated color temperature (CCT) describes the different light sources, which have different distributions of colors which they emit, describing the color temperature as a measurement stated in degrees Kelvin (K). Kelvin is actually a measurement of absolute thermodynamic temperature. We use Kelvin for color temperature as the principle of the color that is emitted by a

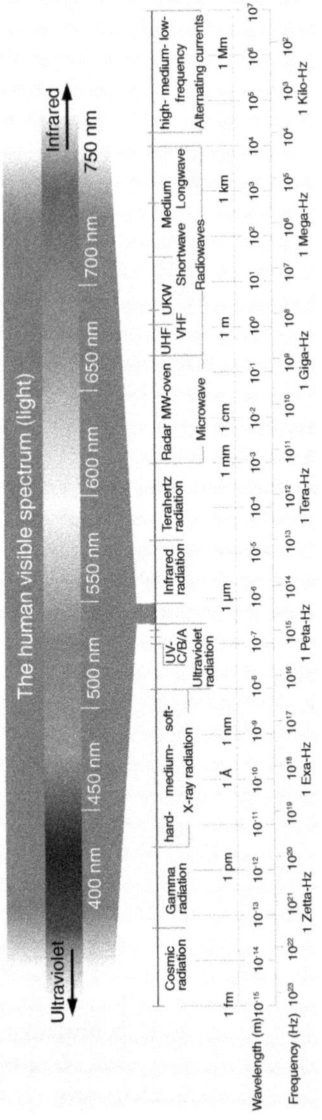

EXPECTATION AND PERCEPTION: 2D IN A 3D WORLD 21

Figure 2.1 The electromagnetic spectrum.
Source: Horst Frank / Phrood / Anony, CC BY-SA 3.0, http://creativecommons.org/licenses/by-sa/3.0/, via Wikimedia Commons.

black body when it is at a specific temperature. Something that has a color temperature of 2,700K will be described as warm, like a typical household incandescent light, while another source above 5,700K will be described as cool. Direct sunlight will be measured at 4,800K, but when trying to emulate "daylight," especially for video recording, a light source of around 5,600K is desirable, though the Society of Motion Picture and Television Engineers (SMPTE) defines reference white as 6,500K.

While light has been around all of our lives; we take it for granted and do not always pay attention to what it is doing. Light is a transverse wave which oscillates in a direction perpendicular to its travel. We know that when we block a light source, we create a shadow. What may be less noticed is that the shadow may have soft or hard edges or be lighter or darker depending on various factors. Learning the patterns of light and how various sources of light will impact the resultant image is the key to calculating how to create a quality image.

In order to predict the characteristics of light, the designer should understand the three main factors that impact it. These components are amplitude, frequency, and wavelength. The amplitude of a light wave is that which is perceived as brightness; the greater the amplitude, the brighter it will be. This is measured in a variety of ways, such as foot-candles, foot-lamberts, candela, lux, and lumens. When looking at a projector's specifications, its brightness is expressed in lumens. Light can be increased by stacking multiple light sources in the same area (try shining two flashlights on the same object and it will be noticeably brighter). This is known as wave propagation. The intensity change is either in-phase coherence which results in a brighter change, or can have darker fringes due to antiphase coherence.

The other two components, which are wavelength and frequency, determine the color of light and the intensity or saturation of that color. Longer wavelengths lead to reds while shorter wavelengths lead to blues and violets. Obviously, what we perceive has additional factors beyond these three components as projection also relies on what the light reflects off or transmits through. This will be more apparent as we look at all variables.

Light radiates equally in all directions. However, that is not productive to how the projection designer requires light to behave in order to produce an image. Manufacturers utilize known methods to guide light to make it useful. These properties include reflection and refraction. Reflection redirects light waves by bouncing them off a surface to send the light in the desired direction. Refraction, on the other hand, bends the light while it transmits through the surface. Together, reflection and refraction allow us to harness light through the use of geometric optics to manipulate it into images.

Reflection is the main property that a projection designer needs to understand as it is the primary means by which the audience will see the projected image. All opaque objects are seen as a result of the property of reflection. The surface texture dictates whether the light will reflect in regular or irregular patterns. In regular patterns, such as a mirror, we understand that light is reflecting in a relatively predictable manner, in that the angle of incidence determines the angle of the reflection. So, when we stand in front of a mirror, we see ourselves, but when we use a mirror in our vehicle, we use it to see off to the side, outside our normal field of view. The mirrored surface reflection is known as specular reflection.

Reflection can also be irregular. A diffuse reflection comes from rough surfaces in which the surface feature height is equal to or greater than the wavelength of incident light. The light is then reflected in random directions. When light is reflected in this manner, the object is able to be seen from a greater field of view. An example of an irregular reflective surface is the page of a book, which is not noticeably less illuminated when being looked at from a multitude of directions.

Conversely, refraction is the altering of light as it passes from one medium to another. For the purposes of the projection designer, the two most common uses of this principle will involve the optics of the display device and the use of a rear projection surface (a surface which is between the audience and projector). Within a projector, refraction is used within lenses to direct the light in a predictable spread and gives us the ability to focus the image.

Different lenses can allow for the focal plane to be variable or fixed in size, depending on the distance from the projector.

Refraction is a tool which can be predictably controlled, similar to reflection. The speed at which the light changes as it passes from one medium to another is observable in two ways. With a prism, white light can be observed to break apart into the various wavelengths, creating a rainbow. Also, if you look at an object submerged in water, it will appear distorted, possibly even dissected, depending on the angle at which you are observing it.

While it may seem impossible, refraction can have reflective properties. This is used both in the optics of a projector's light engine, and in fiber optics (specifically birefringence). The principle known as total internal reflection (TIR) is actually refracting light in a reflective manner. As light encounters a border between two transparent mediums at a critical angle, instead of the normal refraction the light bends in a similar manner to a reflective surface, staying within the medium. The angle for the boundary between air and water is approximately 49 degrees. The angle to achieve this depends on the medium and the wavelength of light.

Both reflection and refraction have common shapes which are utilized to control the light. A flat surface is used for a mirror to reflect with little distortion. A flat surface in refraction can help offset an image when placed at a slight angle from the direction of the incident light. A concave mirror (like a bowl) bends light inwards, which has the effect of magnifying the image near to the surface. This also concentrates the light source within a projector to use as much of the light as possible. Meanwhile, a concave lens will spread the light as it passes from the flat surface through the concave surface. Finally, a convex mirror allows for a wider field of view or will spread light. The convex lens, on the other hand, will bring the light together.

Reflection and refraction are the main ways that we control light for our use. There are three additional principles that the designer will need to understand as well. The properties of dispersion, scattering, and absorption can have a major impact on how an image is presented. These properties are going to be mostly control issues after the light leaves the projector.

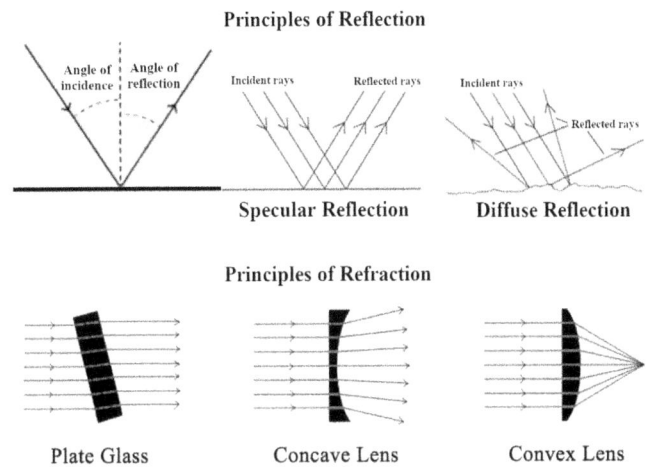

Figure 2.2 Methods of reflection and refraction.

Dispersion is the principle where white light is separated into separate colors as it passes through a transparent material. White light will separate into around six distinct colors. Wavelengths with less energy and lower frequency will slow down less than light with higher energy and higher frequency. The amount of dispersion will depend on the angle at which light enters the medium and how long before it exits. When choosing a material for rear projection, the designer may need to experiment to find whether the surface may disperse the light, causing an imperfect image. Unstructured light that is encountering interference can be seen in everyday phenomena such as the iridescence of a soap bubble.

Another means of altering light in nature (and artificial substitutions) is through scattering. The lighting designer may be using this principle in order to enhance their look by allowing the audience to see the beams of light. They will use an atmospheric effect such as a water-based haze to allow the beams to be seen. There are two distinct methods of light scatter: Rayleigh scattering and Mie scattering.

In Rayleigh scattering, the wavelength of light is much greater than the particles it comes into contact with. This is what gives us a blue sky. Blue wavelengths are shorter than other wavelengths, and are scattered with greater frequency, making them more visible. This type of scattering is less impactful for the projection designer, with the exception of outdoor productions when there is some sunlight. At sunrise and sunset, more red wavelengths can be perceived due to the fact that light is passing through more of the atmosphere, scattering the blues to a greater extent so that they are less perceived and replaced with those with longer wavelengths. Even with an extreme distance between the projector and projection surface, the amount of scattering is so minimal that it is imperceptible.

On the other hand, Mie scattering (where particles in the path of light are similar or slightly larger in size to the wavelength) increases the scattering of more wavelengths, making all of them more visible. This equality in scattering incident light is what allows us to perceive milk and clouds as white. In humid climates, the sky will be more of a pastel blue than arid regions. So, when the lighting designer chooses to use atmospheric effects to emphasize the beams of light from their fixtures, it will have a direct impact on the projection as well.

Finally, not all wavelengths of light may be transmitted to the eyes of the audience. When we see a cherry red sports car, it is because the paint is absorbing most of the visible color spectrum, only reflecting those wavelengths that we will perceive as that cherry red. This does not mean that there are no blues or greens in that which is reflected to the observer; they may be present, but at an almost imperceptible level. However, the amount of their presence may be greater than another red pigment – for argument's sake, candy apple red.

BIOMECHANICS – HOW WE SEE

Regardless of how light works, if it cannot react with our brain, it makes no difference. As we really do not think about how light works, the only thought that is usually given to how we see is whether or not it is in focus and if we need something to correct that focus. There is so much work that our eyes and brain do in order to take in light and interpret its meaning. Biological adaptations to our environment helped our ancestors survive by being able to distinguish predators and how far away they are, whether or not something could potentially be poisonous, and eventually to recognize beauty (hey, mental health is just as important).

When the projection designer recognizes how we see, they can utilize that to manipulate their designs. This allows for being able to compensate for not having desired equipment (underpowered projectors, lower resolution, etc.) by tweaking the design for how the audience will perceive the end result. It can help the designer predict what equipment might be needed for a specialized effect. Just like a magician using misdirection to pull off an illusion, knowing how we see and how our eyes and brain interpret that information truly makes the projection designer a magician in their own right.

First and foremost in how we see are our eyes; without them, we do not have a method of taking in light. Essentially, all animals are able to perceive some form of light through photoreceptive cells, but we really only need to understand the unique way that humans see. Our eyes manipulate light, using the same refractive methods discussed previously. Light enters the eye through the cornea, which is a convex lens containing a transparent liquid. Light then passes through the pupil, the black center when looking at an eye. The colored muscles surrounding the pupil are the iris, which control the amount of light entering the eye by adjusting the size of the pupil. Immediately after light passes through the pupil, our eyes have a lens, which is able to be dynamically shaped by muscles surrounding it. This allows the light to be actively redirected for a clear view of objects at varying distances. After passing through the lens, light passes through the largest portion of the eye, the vitreous humor, which is a transparent jelly-like substance, to the retina, which is the inside wall of the eye. This is where the photochemical receptors, known as rods and cones, are located. These two types of photoreceptors give us a range of

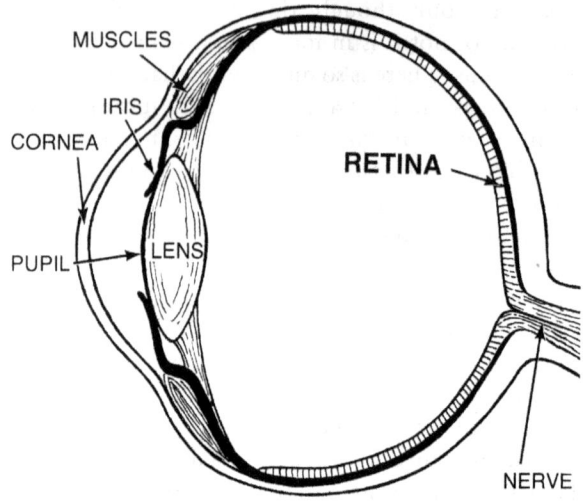

Figure 2.3 Anatomy of the eye.
Source: Pearson Scott Foresman; image in public domain.

sight. The rod cells detect varying levels of light, and are primarily useful in low-light conditions. We have three types of cone cells, which are able to detect red, green, or blue wavelengths of light, giving us the ability to interpret a range of colors. Low-light vision is known as scoptic vision, which does not excite the cone cells, giving what we see primarily as monochromatic. Meanwhile, photopic vision, where the cone cells are active (bright and daylight settings) allows humans to perceive color best. Data from the rods and cones is transmitted down the optic nerve to the brain in order for the information to be interpreted.

The main purpose of the optical structures of the eye is to focus the light and optimize the intensity. As just described, the light must pass through several transparent structures to reach the photoreceptors through various means of refraction (primarily the cornea). Images from distant objects require little refraction, meaning that the lens is relatively flat. For objects that are closer to the observer, the lens must be much rounder; otherwise, the focal point is behind the retina. There are multiple refractive abnormalities which may alter how an individual might see, such as cataracts or astigmatism, but that is outside the concern of the designer.

Interpretation of the light getting to our brain is critical. Our brains have to do interpreting even if there is nothing abnormal. The physics of how light enters the eye inverts the physical world, a principle discovered and replicated in the camera obscura. The first thing the brain then must do is invert and reverse the data so that up and down, left and right coincide with the rest of the senses.

BOX 2.1 THE CAMERA OBSCURA

The camera obscura, meaning "dark chamber" in Latin, was an ancient discovery that in a darkened room with a pinhole to a well-lit exterior, the outer image was projected to the opposite wall, but was inverted. This principle, recreated in a box with a lens, allowing light to fall onto a photoreceptive plate, was the introduction to the modern camera.

The brain's primary function of the optic information is to translate that information into something useful. However, sometimes it can be tricked into something that isn't real, which is exactly what the projection designer needs to do. Commonly, this is known as optical illusions. We will discuss a range of ways to trick the eye in a bit.

How much light do we need to see? This is going to be a continuing question throughout the design process. We should look to see how our body deals with light coming at us. The pupillary reflex is an automatic response to regulate light intensity. The iris sphincter regulates the amount of light entering the eye to keep it within the dynamic range of the photoreceptors. The minimum intensity of a stimulus required to produce a response is known as the threshold. An adequate stimulus produces a response with the lowest threshold. Our brain uses this threshold for action potential and perception of the surroundings (what we need in fight or flight situations). Another quality that our brain uses is saturation, or the maximum intensity of stimulus that produces a response. Finally, the dynamic range, which is the range of intensities that produce responses from the receptor, is the difference between the threshold and saturation. The dynamic range, along with the threshold and saturation, will be different from individual to individual. Certain baselines have been set by professional organizations such as the Audiovisual and Integrated Experience Association (AVIXA) that have recommendations for intensity for various applications.

Just as the set designer does not need to know the background of linear measurements, but needs to understand what that measurement means in context with the design, the projection designer will need to be familiar with the amount of light necessary to realize their design. The answer to how much light we need to see is not really about how much light is being produced, but how much is being received by the audience. There are a lot of factors that impact the amount of light between the source and the observer. As previously stated, light radiates equally in all directions. It is up to how we manipulate light to its intended path which will help determine how much initial light is required.

Even though we use reflectors and lenses to align the light waves to a particular direction, the light will still radiate from the source and spread out. To understand how much light will spread out, we apply what is known as the inverse square law. This means that when the light travels double the distance, it will be one-quarter as bright but covering four times the area.

In addition to the intensity of white light, we need to consider how bright colors are represented as well. As we stated, our photoreceptors are limited to three colors, but we "see" a wide spectrum of colors. Our brains are able to interpret all of these colors by combining the three primary colors of light (red, green, and blue) into a multitude of colors. As this is subjective to the individual, colors are also measured. Quantification of colors has commonly been measured with the International Commission for Illumination (CIE: French – La Commission internationale de l'éclairage) chart. These chromaticity diagrams illustrate a range of colors, representing the average range of human vision as it pertains to the limits of each color we see, being divided from white (at the center). This is only a 2D representation of the colors available to be seen and is not wholly accurate. The diagrams do, however, provide manufacturers with the ability to show the range of colors that a display can produce in connection to what the average person can see, especially in correlation to television standards.

BOX 2.2 MEASURING ILLUMINATION

Different displays may use different measurements of illumination, but they can all be calculated to one type for the designer to understand how much light is necessary for the design.

The 26th General Conference on Weights and Measures (CGPM: French – Conférence Générale des Poids et Mesures), being the supreme authority for the International Bureau of Weights and Measures, redefined photometric units in 2018 (as has also been done in the past). The measurement of light is based on the candela (from the Latin for candle), or 1 lumen = 1 candela * sr (steradian or square radian, the unit in the International System of Units (SI) for a solid angle, i.e., the tip of a cone). This is roughly the amount of light produced by a single wax candle.

Luminous flux (lumen) is weighted for the human eye, measuring wavelengths that are perceptible over time, as opposed to all light produced (radiant flux). The weighting of the wavelengths is due to the human eye not perceiving different wavelengths equally. Those wavelengths that are perceptible are given a weighted sum. As this light is measured in all directions, being focused into a narrow beam will increase the number of candelas within that beam.

Other measurements may be used to describe brightness. The nit (from the Latin *nitere* – to shine) is the number of candela per square meter. A nit was originally measured as the light produced by a single whale oil candle. This is most often seen as the measurement of direct-view displays.

In addition, lux may be used, which is one lumen per square meter. This helps describe the amount of light hitting the surface from the light source. This measurement will help in determining the actual contrast ratio as it can measure the amount of ambient

> light on the surface. If using a light meter for the United States customary units (non-SI), then it may record foot-candles. This is measured in lumens per square foot. For conversion's sake, one foot-candle is equal to approximately 10.76 lux.

COMMON EXPERIENCE

Just as light behaves in a predictable manner and our eyes roughly interpret light in the same manner, we have a relatively common experience in how we see the world. This is what the designer will count on when planning how visuals will look. Obviously, there will be some cultural differences in how we perceive different material, but just seeing an object will be relatively the same. The design process will utilize best practices, knowing that there will be some outliers that will perceive the presentation differently.

All of our senses can become fatigued with regular use. When we put on our clothes in the morning, it doesn't take long before we are ignoring their presence (as long as they are comfortable). The same happens when we hear certain sounds – we eventually stop acknowledging them. More often, after being subject to loud sounds for a period of time, then walking into a quiet area, an individual will likely still be talking at the level of the loud environment. The same happens with our eyes, both with intensity and color. When we walk inside from a bright sunny day, it takes a while before we are able to see clearly as the rods were "bleached" by the intense sunlight and take time to recover. It is hypothesized that pirates wore an eyepatch over one eye as a preparation for boarding another vessel so that they would have one eye adjusted to the dark when they went into the hold, preventing them from being a blind target as their eyes adjusted.

In addition to having our rods being fatigued from bright light, our cones can also be fatigued from seeing too much of a single color, especially if it is a bright source. One type of optical illusion utilizes this exact process. The instruction will be to stare at a single point for a certain amount of time, and then when instructed to look at a plain white surface, the observer will see complementary colors in place of where the previous colors had been, even though the new surface has no image at all. How long this lasts can vary, based on numerous factors. Meanwhile, a projection technician who is trying to calibrate the colors on multiple displays may end up working against themselves if they do not look away often enough. More often, they should rely on calibration equipment which will not fatigue the same way as their eyes.

How well we see and are able to perceive what we see by being able to discriminate the fine details of a scene is known as acuity. There are three main types of acuity: spatial, temporal, and spectral. Understanding what the audience will be able to discern may impact what is designed. Digital displays are in effect the modern version of the artistic technique known as pointillism. This technique of painting used small distinct dots of color in which the applied patterns form an image. How effectively the display works depends on how well it works with human visual acuity.

Most of us will think of spatial acuity first and foremost as we will have at some time in our lives been in front of a Snellen eye chart to determine whether we need corrective lenses. Spatial acuity is a function of location. The periphery of our sight is lower than at the fovea, which is the portion of the retina that contains a higher concentration of rods and cones. Spatial acuity is also a function of brightness: as illumination increases, the ability to resolve the object of focus also increases. This will be important to remember when we later discuss contrast ratio, which is the ratio of light to dark.

Temporal acuity is the ability to distinguish visual events in time. This is what allows us to see motion pictures and video. The critical fusion frequency (CFF) is when a flashing light appears to be continuous rather than repetitive. If the frame rate is too slow, then the individual images appear to flicker. Assumedly, this flicker is what gave moving pictures the nickname of "flicks." As rods are more sensitive to intensity of light, the CFF is lower than that which can be perceived in the cones.

The ability to distinguish differences in the wavelength of stimuli is known as spectral acuity. Far-red wavelengths can excite cone receptors without rods. For this reason, red light is used to light critical applications which require actions of high acuity without bleaching rods. This allows for movement into dimly lit areas without the loss of sensitivity of dark adaptation. Common uses of this are dark rooms for developing photographic film where red light is used for being able to keep light levels low while allowing for the acuity needed to process the film.

Many of us who have spent time on social media may have had to experience another means of our interpretation of what we see as incorrect. The debate first started in 2015 with a dress that was seen either as white and gold or blue and black. Thinking that these are dramatically different interpretations would be correct, but has a lot to do with the quality of the image and how we are experiencing it (both interpretations can be seen on the same display by different people). It was reintroduced in 2018 with a shoe being pink with white laces or gray with blue laces. Not knowing what either looked like without seeing the original allows our brains to look to fill in the gaps of information. In reality, the color debate started even before the advent of social media with the color of Han Solo's coat for the planet Hoth (*The Empire Strikes Back*) being either blue or brown. Again, very different colors, and it was such a debate that the toy manufacturer Hasbro released action figures with both colored coats. In reality, the coat is brown and our eyes were tricked based on exposure and the background colors.

COMMON MISCONCEPTIONS

It could be argued that all issues with a potential design would be just left up to individual perception, but this falls short of complete explanation. Understanding what we see and how we see it can allow the designer to correct for some misperceptions of the design. However, there are still many more misconceptions on what creates a good design, even beyond the content. This usually stems from lack of experience.

The first misconception is that projection design is just "fancy backdrops." As will be explored in greater detail throughout this text, the reality is that media design is a dynamic and evolving field that goes far beyond static backgrounds. It is a versatile tool which is capable of creating dynamic environments, visual effects, and immersive atmospheres.

It may be misunderstood that projection design is out of reach for anyone but those putting on large, lavish productions. Projection is ultimately scalable, allowing it to adapt to a variety of productions. Often, the impact of creating intimate and immersive experiences is more often achieved in a close setting. Access to the required technology is increasingly improving for a greater number of theatrical endeavors.

However, one challenge that a projection designer will face is time. For smaller productions in particular, a director or producer will often consider using video as part of the show well into development. There are many reasons that projection is considered an option midway through a production. Whatever the reasons, the lack of time will have an impact on the end result. There are different processes that all have time constraints, and concessions will need to be made in order to handle them in order to have a finished product.

It is widely believed that digital content creation is virtually effortless and expedient. There are some truths and some falsehoods here. If we are to compare a projected backdrop with a painted one, it is, in a way, true to state that the digital image is quicker to put in a show. Even if we compare a projected backdrop or a traditional backdrop from the same scenic supply house, the delivery of a digital file is almost instantaneous as compared to shipping a traditional drop. However, hanging a drop can be considerably less taxing than setting up a projection system.

Generally speaking, a director who is bringing in a projection designer late into the production is not looking for someone to simply add content from a rental house, but is hoping for something custom. This is where time will be limited. What determines the length of time required? There are many factors. Let's suppose that director and designer are able to quickly come to a consensus on what the design should look like (sticking with projected backdrops only). The designer will require time to create the imagery. The amount of time required will vary, depending on detail and final resolution. If it is animated at all, that will add additional time. Many productions are set in multiple locations and periods (time of day, seasons, etc.) and will need more than one look. Each scene will take time to create and render. Then the designer may want to create transitions between each look, as opposed to fading to black between images. Those transitions will also take time.

Another factor to consider is the time for setting up and configuring the video equipment. Unfortunately, inexperienced production staff will go by their experience in their personal lives in judging the time it takes to set up a home entertainment system and may liken that to setting up projection for their show. There will be considerable differences. Technical departments are often competing for slots in the schedule to finalize their creative processes. Projection also needs many of the same time slots, which will compete with other departments. Installation of equipment will fall with other elements of load-in; however, and as with lighting, projection requires dark time to focus. This may directly compete with the ability of other departments to complete their tasks. When working with single projectors, this can be expedited. When working with multiple projectors that are part of a larger system (aligned together to create a large image or with the images on top of one another for brightness), this can become very time-consuming, taking many hours to get it right.

Just as it takes other departments time to create cues, it can take considerable time to create cues for projection. In earlier productions, this may have been as simple as finding the right points on a video cassette tape and pushing play at the appropriate time. With almost all projection cues being computer-based, programming time is needed before technical rehearsals can take place. This is to take into account working with media servers which are often doing some rendering in a live environment. Depending on the nature of the cue and playback system, the time this takes can vary and needs to be accounted for.

This leads to the next misconception – cost and complexity. Time is money, and money is often a concern when deciding to use projection on a show. If a production company does not already own all of their equipment, they may have sticker shock when they decide on adding projection. Going back to the predeveloped video backgrounds, many content companies will rent the equipment necessary, including projectors and playback devices. This may not suffice for all productions, which may seem obvious for those looking for more in their design. The bigger the projection, the greater the distance between source and display, and additional complexity of design are all factors which will raise the cost.

Projection designs can certainly be very expensive and quite complex, but they can also be achievable for those with less experience and tight budgetary concerns. As with any production, some equipment may need to be rented or purchased used. Also there is a variety of software which can be used and has become relatively inexpensive (some include rental options) and user-friendly. Introduction to projection design is something that is available to productions of almost any size and budget.

Large-scale projection requires powerful projectors to compete with ambient light, which is light from other sources than the projector. Beyond the high price associated with each projector (likely the design will require multiple projectors to be bright enough), this will require a lot of power. While this is not going to be at the top of the list of planning when designing a show, it can certainly have an impact if it is ignored. A typical projector used in a meeting room may only need to be 3,000–5,000 lumens, and can run on a typical business power outlet (in North America) which offers power of 15–20 amps at 115 volts. Meanwhile, a single high-power projector which offers 20,000 lumens or more will, with few exceptions, require 20–30 amps at 208 volts. This can be problematic from the start as it is uncommon for many performance spaces to have the required amount of power available, without some additional equipment. So, it could be considered that the final misconception is that the required amount of power will be available for projection to be used.

WHAT MAKES IT RIGHT?

The desire to add projection to a production can be fraught with challenges. Questions may arise such as "Is there a right way to do it?" or "Is there a way to measure the quality of a projection design?" Consider that there is not yet a Tony Award for projection design, so that could lead to the answer being "no." In all honesty, the true measure of the success of a design is personal. What expectation was there at the beginning of the process and did the results meet those expectations? That is the true measure.

That being said, there are methods to make improvements. We will discuss best practices throughout this text; we will look at some things that will make a presentation as good as it can be. Not every production will have the means to accomplish what others will. That should seem obvious, but many directors and producers insist that their lower-budget production should be on the scale of well-funded productions. While there is no issue with striving for the best, there is the risk of cutting the wrong corners, even to the point of making something unsafe.

To get started, we will talk about illuminance, or the amount of light being used to project the image. As the human eye is primarily attuned to changes in light and dark, this is the primary concern for the designer in presentation. A display will likely notate the contrast ratio as part of its specifications. This can vary widely depending on how it is measured. Some manufacturers will attempt to increase the contrast ratio through various means of reducing the light. No matter, though, that is only how much is being created by the display, in actuality the contrast ratio needs to also be measured after the light has reflected off the projection surface as there will be inevitable ambient light. This leads to one of the biggest misconceptions in a projected image – the ability to project black.

Black is simply the absence of light. If a portion of an image needs to be black, then as much light as possible needs to be kept from hitting that part of the display surface, or the surrounding light needs to be to a level that our eyes adjust to not seeing that portion of the image. For instance, if a projector has a contrast ratio of 400:1, then for every 400 lumens of white light, there is 1 lumen projected in areas that are intended to be black. This means that if you have a 10,000-lumen projector, the areas intended to be black in the image will actually be 25 lumens projected to the surface. This level of light projected in areas intended to have no light is known as video black. So, instead of appearing black, those areas appear as a very dark gray on a white projection surface. This can have a profound effect on the lighting designer when they intend to have a complete blackout.

Ambient light, sometimes known as light pollution, is light from other sources than the intended light of the projector falling onto the projection surface. This is competing light that is not intentionally being transmitted to the surface, such as a spotlight that is reflecting off the floor, hitting the projection surface. While light waves can essentially cancel each other out if they are at specific opposing frequencies at a specific location (in a situation such as double-slit diffraction where some spots will destructively interfere with one another), most often they amplify as constructive interference (in-phase coherence). Even small amounts of light falling on a surface, adding to the video black, can have devastating consequences on the visibility of the projected image. In particular, darker images require much more control of ambient light for the audience to discern details than an image with brighter colors.

In order to get a better image, the projection designer has limited choices: control light hitting the surface to the projector solely, increase the amount of light of the display, or change the content to be less impacted by ambient light. The next misconception usually is due to not knowing how to control these elements. The technique is often reactionary; altering settings in the display is often the first choice in order to make the image more visible, but is actually the result of poor planning. With a direct-view display, such as a television or monitor, with which most people have some familiarity, settings such as brightness and contrast will impact the image differently than with a projector.

In some direct-view displays, such as an LED wall, contrast can be much better as the portions requiring the image to be black may have their light source completely off, in addition to the light-absorbing material between each light source. However, some direct-view sources (such as a video monitor) cannot achieve true black as they have a global light source and still emanate video black. In this instance, an adjustment to the brightness raises the black levels (reducing contrast), which does not really improve the visibility of the darker images. An adjustment to the contrast ratio of the display will adjust the ratio of light to dark, deepening the darker portions of the image. The greater the ratio, the better the image; and it may be necessary in situations that have less control of surrounding light to adjust these settings. The colors displayed will likely be altered as adjustments are made to brightness and contrast ratio.

> **BOX 2.3 SUMMARY**
> - Light is predictable and understanding it can lead to predictable results.
> - Amplitude, wavelength, and frequency are the three components which characterize light.
> - There are five principles in how light is altered. These are reflection, refraction, dispersion, scattering, and absorption.
> - The eye is an optical instrument; it uses the cornea and lens to focus light on the retina to stimulate photoreceptors.
> - Cones and rods are the photoreceptors that detect the frequency (color) and intensity (brightness) of light.
> - The amount of light produced is measured in lumens.
> - There are three types of visual acuity which allow us to discriminate details: spatial, temporal, and spectral.
> - Three primary misconceptions in regard to projection design are time, money, and the amount of power required – though others exist as well.
> - Optimizing brightness and contrast ratio of the display, along with choosing content which fits within the limitations, can all improve the overall quality of the design.
> - Success of a design comes down to expectations.

DESIGN ELEMENTS AND WORKFLOW

With a basic fundamental idea of how projection has been used in the past as well as the way it might look, based on the principles of light and perception, it is time to look at the design process. In many ways, the process of designing for video will be very similar to other design processes. In some ways, as a completely different medium, the processes will diverge considerably from how other designs are realized. This, of course, may sound very rudimentary in concept as all forms of design will have their similarities and differences. The challenge is in how they are considered, especially when timelines of production needs are to be met and the collaboration of other design elements.

This is where the designer will need to be able to find a workflow to suit the production at hand. Some elements may be unknown, regardless of the experience from prior productions. This often comes from adding some special element, such as interactive projection or putting video into a moving scenic element where wireless transmission is required. Making sure that experimentation and troubleshooting time is worked into the workflow is essential. As always, communication is the most crucial element in the design process. Without it, the most amazing design can come crumbling down during the staging of the production.

Design is "show-centric." The same designer can create completely different elements for the same show, depending on how that show is to be presented. It will be something broad enough for creative freedom and interpretation, but narrow enough for ultimate manageability. It is the ability to move from concrete observations to highly abstract thinking (and then back). It is a process of divergence (brainstorming) and convergence (decision-making).

Design fulfills many purposes as will be discovered along the way. It allows the director (through the designers) to amplify what is good and remove what is bad, focusing on emotions of the story. It allows show elements to be taken to the extreme and to question assumptions. It creates analogies. It allows the audience to focus on a single element or break it into pieces. Various resources are leveraged to accomplish these goals.

THE PURPOSE OF THE DESIGN

What is the purpose of the design? This may seem like a ridiculous question. The two main purposes that the projection designer must meet are to complete the vision of the director and to support the production as a whole. After all, at the core of design is how the artistic vision drives the creation of compelling content. So designers must align their work with the overarching narrative, ensuring a seamless integration with the intended message. How that is accomplished can be answered in innumerable ways. What is the style of the production? Even a show whose main intent is spectacle can have a projection design support or overshadow the rest of the show elements.

During the early meetings with the director, it is good to know if they understand how projection and other video elements work. It is good to start with realistic expectations. Knowing the baseline that the director has in what they are requesting will give the designer the ability to not have to lead with "no" or unrealistic "yes" answers, but the ability to offer suggestions throughout the process of what can be accomplished within the constraints of physics and budget. It will also be helpful to assist the director in envisioning the final product throughout the design process.

This will help them focus on the second element, supporting the production as a whole.

Video projection can be quite enticing to play with in many ways. It can add cinematic qualities that other live elements can only allude to. It can rapidly allow the director to change location. It can give performers an extra bit of magic. On the other hand, it can easily upstage the rest of the performance. If done wrong, it can take away the suspension of disbelief instead of enhancing it. Even when the production is purely spectacle, such as a circus production, the video elements can distract the audience from the performance, though the design elements may still support the rest of the show.

SCRIPT ANALYSIS

While not every production will be a play or a musical with a script to follow, there will likely be a theme with some sort of outline. The following questions and suggestions will be relevant in some form, whether the production is a traditional play or musical or some other form of performance art. By delving into the intricacies of the script or other outline, the projection designer can create a holistic and memorable experience. We will discuss this section as if it is a traditional show, with a script. Those who are designing for other productions can utilize any outlines or other thematic description of their show in a similar fashion.

The designer may or may not have a meeting with the director prior to receiving the script. Assuming that the projection designer is brought in from the beginning, then let us also assume that the director wishes all of the designers to familiarize themselves with the script first. Script analysis is a foundational step in the design process, allowing designers the ability to align the elements with the storytelling aspects of the production. If the performance will be a revival of a previous show, the first reading may end up evoking memories of what has already been produced. The director may be looking to recreate a similar production with a fresh take or they may be looking to completely diverge from something done in the past (consider the many variations of Shakespeare's works, told in classic style or modernized). During the first reading, the designer may feel inspired by what they have seen previously, thinking of how they would recreate closely or considering how they could improve on what they found as shortcomings. However, the designer should also keep in mind that the director may be using a familiar script, but will have something else in mind than what has previously been produced. Keeping an open mind and attempting to read the script as if it were completely unknown should be a priority.

This is obviously easier if the script is completely unfamiliar. This gives the designer the opportunity to experience the play as an audience member, getting to meet the characters, learning the story and themes, feeling the emotion for the very first time. This allows for getting an overall understanding without preconceived notions. This is not the time to be focusing on the projection design per se. This is the opportunity to feel what the characters feel and see what they see. The designer may choose to take notes of something extraordinary that really impacts them (not necessarily something visual, it can be a feeling or something that evokes another sense). Take note of things that the dramaturgist may have specified for settings and time period. Clues to the place and time as well as the general feeling of the production are often gathered directly from the dialogue. This can be written with accents which are discernible in how the dialogue is written as opposed to a description of how words are said. Again, the director could have a different idea of how the production will go, so the designer truly needs to keep an open mind.

BOX 3.1 SCRIPT ANALYSIS QUESTIONS

- What style is the production?
- What is the mood of the production?
- Does there seem to be an overall message to be conveyed?
- Where does the production take place? Are there multiple locations; interior or exterior?
- What time period is the production to represent? Are there multiple time periods?

- How does the production transition between scenes? Are there transitions within a scene?
- Are there required effects listed in the script? Is there a listed source?
- Are there required visuals listed in the script? Will this require special permission?
- Are there particular visuals described by the characters in dialogue?
- Do any visuals play a part in advancing the action?
- Do visuals need to set the scene before action begins?
- Are there subtle ideas which could be enhanced with visuals?
- Are there particular cultural references that might be sensitive?

While the first read is to get the general understanding of the play, the second read will be where the designer needs to take a much more in-depth look at the details. This is where the ideas will begin to blossom. The second read will be for analyzing specific points where visuals are directly and indirectly called for. Consider any description of the use of technology in the play where a practical element such as a television or projector (even a slide projector) is called for. Look for special effects which could be difficult to produce traditionally, such as fire. Note supportive elements which might enhance the scene, such as being able to see a storm outside of the window in the set as opposed to just flashes of light simulating lightning. The designer should envision themselves in the place of the characters, what might be mentioned in dialogue that might be seen in the mind's eye and could be shown to the audience. How will visuals impact the story and how the audience experiences it? This is the time to ask specific questions of what will be seen, but not yet how they will be seen. This is still the discovery process, as gathering ideas to discuss with the director is the key motivation.

The script analysis should contain an understanding of the overall narrative. The designer needs to understand the thematic elements and be able to identify key moments that warrant visual emphasis. A study of the script may identify visuals that could be created to reinforce character traits, plot twists, pivotal moments, and emotional climaxes. In addition, designers must be aware of historical and cultural contexts in regard to time periods, societal norms, and other references embedded in the script.

Once the designer has finished the second read-through and determined what visuals are required as well as the visuals which could enhance the storytelling, then it is time to organize these thoughts into a useful presentation. Begin in order, with an outline of the entire production, noting specific details, such as the page number and where a cue would need to begin and how long it would persist. Other details to be identified are things like whether the image will appear on a practical device (such as a television or computer screen) or be projected on an unrelated object. Making note of the subtlety or boldness of the image may be helpful as well. An understanding of the tone and atmosphere of particular scenes will help align visuals with the intended emotional impact. This all requires sensitivity to the script's nuances and a keen understanding of the director's vision.

Making a simple chart containing all of this information (cue numbers will not be necessary at this point, though numerical listing is an efficient means for communication) is an important conclusion to the script analysis. This will prepare the designer for the first meeting with the director and other designers.

The process should include breaking down the elements in order to understand the spatial dynamics. The designer needs to consider the physical layout of space, including sightlines and potential areas for equipment as well as the ultimate scale of the video elements. Paramount to these proposals will be consideration of available technology, budget constraints, and other logistic challenges. This will be the culmination of the technical feasibility of the proposed design.

PROPOSAL AND COLLABORATION

Prior to meeting with the director and the rest of the team, the designer may need to visually organize their thoughts. This began with the idea chart, but this likely will not be enough to truly express what the vision will be for the production. Since the assumption is that the director has not yet shared their vision, the amount of work prior to the first meeting will have its limitations. There is little point in sketching out each and every visual, especially if it can be easily described, as the designer will be unsure if the director is going to go in a different direction. Thus, visual communication should still be rudimentary. In later meetings more detailed visuals should be used.

Every director is going to have their own style. They may request the ideas from the designers before expressing their own vision, or, more likely, they may spell out how they want the production to appear before hearing if any of the design ideas mesh with that. It is also important to come to an understanding of how the director feels that digital media fits into the production. This is when the design team needs to focus on their ability to listen and ask relevant questions about the director's ideas. Preparation of open-ended questions will be advantageous. Are there specific thematic elements that the director wants to focus on? Are the requests even something that needs to be projected? Is what is being requested something that can work within the scope of the time and budget? Are the ideas of the other designers something that will be compatible with digital media or will there need to be significant compromises? This is the opportunity to gather information in order to have a solid understanding of the expectations of the director and possibly the other departments. Listening is a crucial skill in allowing short answers to build into necessary detail. Attention should also be directed towards the non-verbal and emotional tone of answers. It is good to repeat back the director's intents in order to elicit more insight as to the vision of the production. The next meeting will be the opportunity to share that understanding.

After that initial meeting, the projection designer should have a solid foundation of where the project is headed. They should be able to identify how the show should look and feel in its final form. They should have an understanding of which departments they will need to begin collaboration with so that the designs will be able to work together. It is time to go through the idea chart to see what fits and what needs to be adjusted and possibly added. How do those original ideas flow with the ideas of the director?

Brainstorming is a term that we are all familiar with. It is a process which we may not all really think about or use. It taps into a broad body of knowledge and creativity that the analytic mind may have passed by. This is an essential skill to use after becoming familiar with the script and learning how the director envisions the production. This may be a process that the designer chooses to work on with the director immediately after the initial meeting, to further harmonious collaboration, or to do completely on their own. It is good to remember that brainstorming is a method of promoting openness and generating a lot of wildly creative ideas. This is still not the time to be looking for technical solutions, immediate feasibility, or any other limitations. That will be for a later brainstorming session. However, the designer should have creative confidence. They should start with big ideas and act on them. Ultimately, it may be easier to pare down ideas than elevate something too simplistic.

Now is the time to cluster similar ideas into groups in order to refine the design concept. The designer must make decisions from the previous brainstorming and how it fills the needs identified in the original idea chart. This will be compiled to deliver at the next meeting. There are different approaches the designer may utilize in communicating their ideas in the subsequent production meetings as well as individual meetings with the director and other designers. One or all of them could be used, depending on the type of production. Common methods utilized are mood boards, storyboards, and action charts. These are the early communication steps before content is created or collected for the show.

Overall, the design process should start at the end (how the audience will experience it) and work backwards, though not necessarily in one straight path. If the end goal is not known, the end product will likely be lackluster at best. This is not to say that the end product will be exactly as originally conceived, but it should have some semblance of it. One of the most difficult things to convey is how the production makes the audience feel. How the audience felt will precisely be one of the most discussed topics among themselves after the show is over. These feelings tie into the first communication method that a designer might use.

MOOD BOARD

The mood board will use visual items to help convey the ideas of a scene or the production as a whole. This tool is abstract in nature and is not just limited to projection design at all, but is widely used among creatives. A mood board is a collage of photographs, illustrations, graphic images, text, and other visual items collected to help represent an idea or feeling. Traditionally, this collage would be on a singular surface, such as a poster board, so that the images can be viewed together to get a sense of artistic direction. Modern mood boards may be a collection of images in a computer folder or on a website (Pinterest is a popular site to collect images). Online services can be an easy way to collect the images; however, the display of the images may not be ideal if only viewed as a collection and not grouped into a singular image.

How the designer arranges the images on the mood board can provide an impression as much as the content on the board, toward the feel of the design. Consider if images were collected from magazines and each picture was neatly cut (either squared or neatly around the subject) and arranged, as opposed to torn edges and haphazardly placed. One already has the feel of order and structure, while the other may be chaotic or free-spirited. This can be the opportunity to represent color palettes or stylistic interpretations. Text can be used to show font style (if text will be used in the design) or to help set a tone while describing what is on the page (it is good to have notes accompanying the images). Digital representations can be built in presentation software like PowerPoint or Keynote, in a graphics program, or on a number of online options allowing the collage aspect.

There are merits to both physical and digital versions of the mood board. On a tangible mood board, it may help the designer to also illustrate the physical surfaces of where the projection will be seen. Texture may not be as easily conveyed through just a picture of fabric as opposed to an actual sample of the fabric. However, a digital image may be able to represent what the projected image might look like on that same fabric. A physical board may engage the director where a digital board may not. A digital board may take much less time to prepare than a physical board. In addition, a physical board may be limited to the size of photographs found, whereas a digital board can manipulate individual images to the size needed for presentation (larger objects having greater prominence and value than smaller images). While the designer can share a picture of a physical mood board with the rest of the team online, a purely digital mood board is much easier for distance collaboration. Digital mood boards could potentially include animations (such as GIFs) that would be near impossible on a traditional mood board. They also allow for greater flexibility in change as the project evolves.

A mood board can be practical, illustrating exact images that are to be created (such as a historic site). This may be the case when attempting to tell a realistic production and the designer is looking for specifics, including reproducing newspaper headlines. In the early stages of a production, the designer is more likely to be exploring ideas and the images, while concrete in nature, are representative of what may be part of the design. Most likely, it will be some mixture of knowing specific ideas that the designer will be focused on creating while other images will be to open discussion with the director into how they imagine the production to look.

Ultimately, the mood board fills three primary goals for the designer: inspiration, consistency, and communication. While the designer is putting together the mood board, the images being sought to communicate the ultimate idea will in turn offer additional inspiration to the designer. For instance, if a designer is looking to design a play with a feel of ancient China, themes represented in ancient artwork may help to develop a greater theme to the details in the design. In other words, it can help alleviate "artist's block" when looking to flesh out a design. It will offer a reference point throughout the project to keep the design cohesive and consistent. While it is being created, stay consistent and only add images that support the mood so as not to overcomplicate or clutter the board. Of course, as was the first point of discussion, the mood board is a communication tool between the designer and their team as well as to the director and the rest of the production staff. In the beginning, the designer should plan on multiple mood boards to allow the director to discuss the merits of different ideas, rejecting those that do not fit.

STORYBOARD

Another method of graphically communicating an idea is the storyboard. While the mood board is, even when structured, more free-flowing in how ideas are expressed, the storyboard is more on a directed path. The storyboard will feature a sequence of images and often include some sort of dialogue to introduce the concept portrayed. Similar to a comic strip, they will be broken into individual images which represent critical information or relevant concepts. The images may have some similarities to the mood board in that they may be photographs, illustrations, or other visual representations of how the design might look.

While a storyboard may be most familiar as a tool used by those creating motion pictures, helping to distinguish each shot by a camera, they are very useful for the media designer who can use them to display not only what a single scene may look like, but the transitions between scenes. Physical storyboards have fewer benefits than digital storyboards. Though inexpensive, sharing and updating physical copies are limited, and their creation can be very time-consuming. A digital storyboard can easily replicate elements from frame to frame and can be altered more easily than their traditional counterparts.

A digital storyboard can be created in a number of different ways. Presentation software is highly useful even in simple slide-based forms. It can offer a background which is regulated, but allows for a variety of images and text, and also allows for motion and video. Cloud-based software can allow the designer to work with their team in real time even if they are unable to be in the same workspace. Many of the collaborative software packages even allow drawing with the use of a specialized stylus. With the many benefits, even the ability to avoid costly software, the cost of printing the storyboard for a meeting could be pricey, depending on the quality.

Benefits of storyboarding are many. From the accuracy of showing progression of the design elements to actual linear design, this can be better understood by certain personality types. It is a great tool to accurately present ideas to other stakeholders in the production who may not otherwise be able to visualize how projection is to be used. They offer not just what a scene will look like but, as already mentioned, can also offer how transitions will be made. This can become a major issue later in the production process, so planning transitions early on may be extremely helpful to all involved. The storyboard will also help the designer to explore how to create the visuals, beginning the process of how the design is to be realized. A concise storyboard will also allow greater collaboration early in the production process which may lead to fewer changes throughout the production, saving time and money.

To start a storyboard, the designer needs to create a template that every cell will follow. This is where presentation software can work out well. Going back to the idea chart, the designer provided the basic information which could be contained in each cell. They may already have listed act/scene, page number in the script, possibly a cue number, and the description of what would be created. The latter is what is now to be illustrated, but everything else will be going into a template. If there are notes additional to what can be put on the slide itself, they can be put in the notes section of the slide software which can be shared when printed. That may include additional information about how the projection will be created or specifically the dialogue where the cue will be taken.

Consider that the storyboard may be a form of rewriting the script, in order to tell the story from a different point of view – that of the visuals. The designer should think about the overall premise of the story and tell it in a concise statement. This statement should be brief enough to be stated in one or two sentences, but elaborate enough to evoke an image to anyone who might just be attending a production meeting for the first time and has not yet read the script. Then, in a chronological order, the designer should use the list of key visuals as the start of the storyboard, adding important parameters (such as scene changes) and descriptions which, most importantly, correspond to the concept.

As with scenic design, the projection designer should consider each cell to represent a view of the stage. As opposed to storyboarding for cinema, the audience perspective is not going to change (usually) once they are seated. Keeping a consistent point of view should help the director to understand a sense of scale. In addition, each cell will start very simply as the initial design is to get across the idea of the progression of the design. Detail comes later in the process. It is easy to get distracted by putting in too many details – for the designer this eats up precious time, while for the director and other staff it can take them away from the overall concept. Notes on each cell should be limited, as the visuals should be telling the story. The designer should let the stakeholders add their own notes based on the description given. Important notes may be naming particular characters or scenic elements not otherwise described by the script that could be determined by the heading of the cell.

BOX 3.2 SOME THOUGHTS ON THE USE OF ARTIFICIAL INTELLIGENCE (AI) IN THE DESIGN PROCESS

AI is an extremely powerful tool that is becoming readily available to the world at large. Many different professional designers and their teams are actively using AI in various ways throughout the design process. Thus, delving into the revolutionary integration of AI in projection design, by exploring the transformative impact it can have on creating dynamic and responsive visual experiences in the realm of live entertainment, is paramount. Designers are able to harness intelligent algorithms to explore creative new avenues from concept to adaptive visuals.

While we have discussed the many steps in the conceptual stages of script analysis, AI can assist the designer in this as well. When employed to analyze scripts, AI can begin with identifying key moments, character dynamics, and emotional arcs. The automated analysis streamlines the initial stages of design, assisting the designer's focus on crafting visuals that intricately align with the narrative.

> During the creative design process, the designer may spend hours looking for the right images to use in the concept boards (mood or storyboard). With the use of AI, the designer can instead have images created specifically to showcase their ideas. By using AI to deliver an image, other members of the design team will be able to better visualize the end product as opposed to generic images with descriptors of the focal points.
>
> Meanwhile, during an actual performance, computer vision algorithms can do real-time tracking, keeping visuals in perfect synchronization with performers and scenic elements. It can predict performance patterns in order to simplify complex designs and amplify immersive experiences. Facial recognition is used to enhance personalized involvement. The imagination of the designer can find even more opportunities to use AI in the designs of the future.
>
> As with many forms of AI, using it for projection design proves to have challenges, such as ethical considerations. The future holds many exciting prospects for AI-driven designs through ongoing research. We are on the precipice of a paradigm shift, with unprecedented possibilities for dynamic and responsive visual storytelling. As the algorithms continue to evolve, projection designers stand at the dawn of a new era, where technology and creativity converge to redefine the boundaries of theatrical experiences.

ACTION CHART

The next form of organizing the initial design is the action chart. This is more of an extension to the initial idea chart, but will contain more information. This form of organizing contains more analytical information. It is a logical breakdown of the media scene by scene. This makes it easy to add information provided by the director and during the early process will act as a reference tool for the designer to research visuals. As it is cut and dried, with descriptions as opposed to visuals, it can aid in organization of critical aspects of the production.

Action charts will need to include specifics, such as act and scene for each image or sequence, the page number (definitely the starting point, potentially the duration if not apparent), the location or locations, as well as the day and time. An action chart is often graphically represented as a flowchart. This is a method of logical progression, especially in performance types which may not be as linear in fashion.

Flowchart symbols are perfect for representing progression and change. They provide the flow and logic between concepts and ideas. There are five basic symbols to any flowchart: oval, rectangle, parallelogram, diamond, and arrow. These have standardized meanings which can be used to describe the plot of a play or process of how other productions might work. This may be especially helpful in designing a production with interactive projection or shows which may change regularly, such as a cabaret-style variety show.

The oval is the terminal symbol, representing the start and end of the production. Most often, the design will only have the two ovals, but this is a creative process. In the action chart, the oval can also represent a point of return. It is a stabilizing point.

The rectangle typically represents a process. This is where something is happening. Think of it as a movie screen and to be representing what is shown. For a projection designer, this is the cue. This is where the description of the image should go. For presentation purposes, a thumbnail image would be placed here with the description being in a notes section. The rectangle should then refer back to a more formal version of the idea chart.

The parallelogram is where information is presented in a traditional flowchart. For the projection designer, this would represent outside influences on the system. This is where an interactive design would have a place to represent a performer-controlled device or some other input which can have an effect on the visuals. For a show-controlled system, this could represent where timecode or some other synchronization data is identified.

The diamond is where decisions are made. This is the logical progression of data that is provided. Often, the decision is a simple "yes" or "no." They are not limited to that, especially with a performer-controlled system, where multiple options could be present. For instance, the trigger could be a percentage of a variable. Depending on what percentage is reached (on an applause meter, for instance) a different direction for the next visual would be taken.

Finally, the last basic symbol, the arrow, signifies the direction of the progression. Arrows are singular in direction and can be straight or curved. A straight arrow will lead to the next step in the process, while a curved arrow will be used to return to the same process. For instance, the designer might be using a looping video for a scene, not knowing where the trigger will be to advance to the next visual. To do this, a rectangle would be set to explain the looping video, with an arrow to a diamond which asks the question if the trigger has been activated. If yes, then there would be an arrow to the next rectangle, describing the next visual. If no, then a curved arrow would direct back to the diamond. Each of these arrows would be labeled according to the answer associated with them.

Arrows can get messy if the processes are not flowing linearly. Sometimes the designer will need to connect processes which are not close to one another. In this instance, the use of a circle with a number or other designator adjacent to the two processes can alleviate confusion. For instance, if after a decision one result will reverse several steps, the arrow indicating that decision would connect with a circle and then the corresponding circle will be adjacent to the process that it is going to with an arrow indicating that direction. If the process were to jump a greater distance, say if the action chart is set in scenes and the flow interrupts the normal procession, a pentagon, shaped like home plate in baseball, would be used to jump to another page.

There are numerous other symbols used in flowcharts that can easily be incorporated in an action chart for projection design as the designs become more complicated. By following logical patterns and using them in similar processes to other organizational needs, this can be an invaluable tool for the design team in communicating with each other and the director. This is an extremely helpful tool when working in programming environments that are not based on a timeline or typical cues.

COMMUNICATING IDEAS

Depending on the type of production which the design is for, one of these three forms of communicating design ideas should work. The designer should not have to use more than one, but may choose to use multiple approaches based on the complexity of the design. The complexity does need to be considered based on budget and time, both of which may require simplification of the design.

It will be additionally helpful to prototype these ideas early on. This helps the designer turn these ideas into reality. This will aid the designer in describing the ideas to the team. By doing so, the team may ask new questions, eliciting choices which may lead to the discovery of flaws or additional opportunities. This is the time for "failure," inasmuch as not every idea will make it to the final design. The designer may find that the design may need to be bigger or that the time to create it may need to be altered, especially if some elements will be required for rehearsals.

The prototypes should be considered disposable, so that they are created in a way to convey ideas and not delay the production by making them perfect. This means that the designer may need to find ways of creating low-cost experiments which will not eat away at their final production budget. The "failures" should be quick, cheap, and early in the production process. If there are limited resources, the prototype should not take too much, but leave enough to learn from the experiment and still iterate the ideas.

It is important for the designer to invite honesty and openness from the director and other stakeholders. This is the time to be open to positive and negative feedback and not just the designer to sell their ideas as a matter of fact. The designer needs to ask many questions, especially in order to understand critiques. The ability to adapt on the fly, revise on the go, and be ready to change or eliminate parts of the design ideas is paramount. No matter the feeling of the critique, the designer should remain optimistic. Various constraints due to budget or creative differences will be inevitable and should be viewed as prospects towards unexpected solutions.

Thus, the designer must have the ability to refine their ideas. This process should never be considered truly finished (long-running shows may even have opportunities to refine design elements which keep the show fresh). The designer should consider when, where, and why some of these early steps should be gone through again. When is "too late" to change will vary by production. Thus, every production should be considered as an interactive process: create, share, review, and refine often.

INTERACTIVE OR CHOREOGRAPHED

Projection can be an organic part of any scene. Its fluid nature is the perfect way for a director to skillfully set a scene and draw in an audience in a unique fashion. It can help add magic where even practical effects can be difficult to replicate. For these reasons and many more, a director could easily ask for the world by having all of the projections be interactive. In some ways, this is becoming increasingly easy, to the point that it is replacing some green screen technology in movies and television.

"Is it live, or is it Memorex?" This was an advertisement campaign by the Memorex Corporation when they started manufacturing cassette tapes for the consumer market. In their most famous commercials, Ella Fitzgerald would shatter a wine glass with her voice, while being recorded on a Memorex cassette tape. When the audio was played back, the same results were observed. While we are concerned with video, not necessarily audio, the same sentiment exists in challenging the observer as to what is real and what is not. The best way to accomplish this with video depends on a few variables.

Interactive video means that the content is not fixed by cues and timelines, but can be manipulated in real time, generally by a performer. Many performances seem to be interactive, but in truth, the performers have choreographed movement which is fine-tuned to linear content. Is there a difference between the two styles of content to the audience? A lot of that depends on the quality of the work going into making each design. Both will take time, and both will have their own unique challenges. The designer may have to take a hard stance on which will be better for a production due to the challenges either one of these can have on a production.

The more common of the two will be choreographed "interaction" between performer and video content. Most televised talent shows which have performers and video will follow this course. As seen on those shows, the results can be spectacular. For best results, the scene will be set to music so that the performer is able to have regularly timed auditory cues as to when to do a specific action that follows the video. This makes it difficult for some types of productions, such as drama, where the performer truly feeds off the reaction of the audience, changing the timing of their actions.

The video designer may have to decide whether having video and performer correspond with actions supports the director's vision and the story. Having choreographed video and live action is definitely visually appealing, with a major wow factor, but is seldom necessary for productions that are not just for spectacle. Will there be ample time for creating the visual in time for the performer to have enough time to rehearse with the video to match timing and placement? This will be an extremely important question to ask and requires a succinctly truthful answer. For if there is not enough time to properly rehearse, it will be painfully obvious to the audience. In order to accomplish this, rudimentary visuals may be provided for rehearsals while the refined video will be worked on through the production process. For the best synchronization, some sort of musical soundtrack should be included to provide timing.

Video which works with the performance is an element which can promote liveliness in a production. It helps to accentuate the cinematic qualities many directors desire for their audiences. Because of that allure, the designer may be asked to provide something in this vein. In addition to the performer being choreographed, set pieces and other elements may also move in synchrony. This uses a process of show control: an automation process of using various technologies to operate multiple entertainment control systems simultaneously for a seamless effect. Discussion of some of the methods to accomplish this will be handled in Chapter 7.

Not every production will have the capability of show control. The designer may still simulate a well-choreographed video element by having additional cues triggered by the operator. When utilizing some software, playback of the video can be non-linear as well as containing multiple layers of imagery. For instance, if a character is to jump into a puddle and the director wishes there to be a visual of a splash, a poorly timed choreography could have the splash occur prior to the actor landing. On the other hand, if the operator is watching the action, they could trigger the cue by visually watching the action, allowing the actor to organically time the action. The only caveat to both scenarios is that the performer must jump in a predetermined location for the effect to happen in the correct location.

Specifically due to the challenges of choreographing timing without musical cues and the need to be in a particular location, the director may request that video be interactive. To be interactive, the video will be controlled and manipulated from external sources. Often this involves cameras which allow programs to track movement in the performance area; however, there can be a multitude of sensors. This is an advanced form of projection design which will require a lot of planning and work with other departments in order to make sure it will work correctly.

In order for interactive projection to work, a few different technologies will be necessary. First and foremost, a media program which can accept data from alternate sources and use that data to alter parameters is absolutely required. Preferably, the program also has the ability to generate some content on its own, using tools such as particle generation. Next, a device which can provide external data for manipulation of content is required. Finally, the devices will require suitable connectivity to perform as intended with minimal amount of latency (delay due to processing of information).

BOX 3.3 PARTICLE GENERATION

A particle generator is a tool that is useful in creating organic video effects such as clouds or fire or a flight of arrows. It is a

> method which produces replicated elements which create the illusion through their numbers and animation. These elements may consist of a sprite (a stand-alone computer graphic object that is a 2D bitmap) or a 3D model. Each particle will have variables in how it behaves, including velocity, rotation, scale, and lifespan. These tools have been available in video-editing software for quite some time, but they are now becoming more widely available in media server platforms.

Interactive projection can be as simple as having an object move through a scene, by being layered on top of a background and having the X and Y coordinates manipulated by an outside source, even a computer mouse. There are many common sensors which could be utilized for basic effects. These could include a momentary trigger from a motion detector for lights. Any number of devices could be used to create a contact closure which is useful in starting a video sequence. These simple devices are unlikely to fulfill the desire of the director when asked for this style of projection. Instead, the ask will likely be to appear more magical and hands-free. In order to do this, a system with cameras will likely be used to track movement.

With a large enough budget, there are several companies that currently offer motion-tracking systems that are easily integrated into some high-end media servers. This does not mean that interactive projection is outside the possibilities of those on a tighter budget. With the right program, equipment as simple as a computer webcam can be used. The principles are essentially the same.

The first method of motion tracking is to offer the program targets by which they can interpret what is being seen by the camera. Often, these will be infrared (IR) markers or emitters. The markers are reflective points which are regularly used to define boundaries, such as a set piece, which will allow the program to interpret a projection surface as it moves in space. An IR emitter is similar to a remote control for a television, which emits a series of flashing lights. Each emitter can have its own unique pattern, allowing it to be identifiable in the presence of multiple points. These systems are used primarily by the pre-packaged systems.

What is available to a greater number of designers is to follow objects using blob detection. A blob is an amorphous object which is created by the difference in an image captured by the camera between frames. Generally, the program will be looking at luminance values in the video stream. This means that the lighting in the scene will need to be controlled, as too bright a background can prevent the camera from detecting movement in the foreground. In addition, reflective surfaces or possibly shadows could cause errant detection, so the placement of the camera will be crucial.

Innovative designers have also turned to the gaming industry to utilize some unique control systems in their interactive projection. For a while, the Microsoft Kinect was a sought-after tool. As the video game system had already developed an all-in-one device for detection of people in a 3D space, it was useful especially in dance designs. Its main drawback was the limited depth of field due to the strength of the IR emitter, relegating it to use in smaller spaces. The motion controls in a number of game controllers are also useful. There are even projects using the old control for the Nintendo Wii, allowing for the creation of digital spray paint. They can often be found at pawn shops and used video game stores relatively cheaply, as they lost their popularity.

With the ability to detect motion, various effects can then be created, depending on the variables put into the generative software. The size of a particle effect can be manipulated by the expanse of movement along with the speed of movement. This can now allow the performer the freedom of creating a splash in a puddle at any point in the projection area. In addition, the size of the splash may be varied by how high the performer jumps.

BOX 3.4 SUMMARY

- Design is "show-centric," meaning that the same production can have different designs depending on how it is envisioned.
- The design must complete the vision of the director and support the production as a whole.
- Three methods of communicating design ideas are mood boards, storyboards, and action charts.

- Projected effects that are choreographed to the action of the performers require time to rehearse and generally are timed with music.
- Interactive projection offers many possibilities, including generative effects. These may take more research and development time to get right.

BASIC CONCEPTS

Now that the initial production meetings have passed and the decision is final that projection will be used, it is time to dig into the next steps of the design process. The feel of the production will have been determined; the designer will continue to build their ideas using the communication tools of the mood board, storyboard, or action chart. They utilize a series of design concepts to create the work, in order to be inspiring and appealing, and to effectively communicate the needs of the production. The fundamentals of projection design are concepts which are common to other visual arts. There are the basic elements of line, shape (form), color, pattern, texture, and space. There are also the principles of composition, including harmony, variety, balance, rhythm, dominance, proportion, perspective, and movement.

COMPOSITIONAL ELEMENTS

Each of these compositional components has a dramatic effect on how a design may be perceived. Elements of design are the raw materials to be used in our perception of what we see; visual things are composed of visual elements. The principles of design are the techniques of how the visual elements are arranged to represent the world around us. Combined, the elements and principles serve as a guide to present a visual idea in a strategic and effective

DOI: 10.4324/9781003107019-4

manner. The designer must learn to see prime visual material to be able to perceive the visual features for what they are in order to create a type of blueprint for arranging components into effective and meaningful compositions.

LINE

We shall start with the simplest element, the line. At its most basic form, the line is a connection between two points. However, the line can also become a division between two spaces, and depending on its placement, it can draw the eye to one location over another. The properties of the line can differ. Its weight, which is the thickness, can give the line more prominence, especially when compared to other lines with lesser weight. A line does not need to be straight, but it can be curved or wavy, bringing the idea of motion to a simple object.

A good example for the use of line in art is seen in Figure 4.1. As can be seen, line not only defines the physical structure, but also helps to catch the focus of the viewer. The arch of the bridge

Figure 4.1 *The Ponte Salario* by Hubert Robert.
Source: Image in public domain.

frames some action as well as assisting in describing place (the background visible beyond). Lines in the foreground are more defined, while the background loses definition (more on this as we discuss space).

SHAPE

Moving towards more complex lines is shape or form. Shape is initially a 2D object which may be described as geometric or organic. All visual elements are, in some way a shape. They are defined by boundaries such as lines, color, or negative space (areas between defined shapes). Geometric shapes, often a tool for graphic design, can give a feeling of order regardless of complexity. They are defined through the combination of a specific amount of curves, points, or lines. Most often, they are symmetrical and are easily identifiable. These shapes can be combined to create other recognizable structures, such as a triangle on top of a square to represent a house. Organic shapes are free-flowing and mimic those in nature, but do not need to be photorealistic by any means. The meanings behind a natural shape often mimic the element which they represent through a cultural perspective. Also, there are abstract shapes. These are a subset of organic shapes. These may be simplified versions of an organic shape or may be amorphous and not representing any orderly objects. These forms may have figurative meanings as they can create a unique perspective on an object.

As can be seen in Figure 4.2, both organic and inorganic shapes can be used to create something meaningful. This image, which is the reverse of the portrait painting *Ginevra de' Benci*, utilizes three branches which are symbolic in meaning (culturally relevant). Organic shapes are, of course, the branches. The scroll is an inorganic shape, as it represents something that would be created by humans and would not naturally occur, nor exist in this form.

Shapes are an important means of communicating ideas through the visual process. They may invoke different feelings, depending on the culture that views them. The feelings and behavior of the viewer can be impacted even if the shapes are not directly

BASIC CONCEPTS 67

Figure 4.2 Wreath of Laurel, Palm, and Juniper with a Scroll inscribed Virtutem Forma Decorat [reverse] by Leonardo da Vinci.

Source: Image in public domain.

noticeable. Subtlety of design can often reinforce a message through the use of these elements. Looking through historical images, it can be postulated that artists of the past intentionally put hidden meanings through the use of shape, especially in negative space.

BOX 4.1 USE OF SIGNS

Semiotics is the study of signs and sign-using behavior. Charles Sanders Pierce defined a sign as "something that stands to somebody for something." In other words, a sign is anything that creates meaning to be understood by an intentional audience. According to Pierce, there are three types of signs: an icon, an index, or a symbol.

> An icon is something which we often use to identify a concept without words, such as a floor sign indicating a slip hazard. These icons attempt to illustrate an idea without the use of words to make a concept easily understood in a brief amount of time (road hazards, such as animal presence possibility) or for those who may not be able to understand a written message.
>
> An index does not seek to identify an object, but is more conceptually driven to identify one thing due to the presence of or ties to another. This idea is that the subject sign is associated with a referent, showing evidence of what is to be represented (the most common being smoke to represent the presence of fire or a footprint to show the presence of what created that footprint).
>
> Conversely, a symbol is purely culturally learned, as the sign has no actual resemblance to the signifier (the form of a sign) and signified (the concept or object being represented). Written language is the most common symbolic form of signs as the alphabet and numeric systems have little to do with what they actually represent.
>
> Understanding these symbols without cultural context may be completely impossible, as evidenced by archaeologists attempting to learn symbols created by the ancients compared to theories posited by pseudoscientists for those same symbols. An icon or index may become a symbol over time, such as the image of a floppy disk to be understood as the concept of saving digital work. The study of semiotics may be an extremely interesting topic to the media designer to enhance future designs.

COLOR

Rarely is a design monochromatic. Color may seem to be a simple element, and the most obvious following shape. There may be a misconception that color is an easy design fundamental. Ask any lighting designer, and they will likely tell you how challenging getting the right "look" can be, depending on the color of the light on the subject. Some colors just don't work well together (which could be a specific design choice). Culturally, colors have different meanings which can influence the viewer.

There are two physical means of colors, that of light and that of pigment. Each is said to have three primary colors, by which other colors are created. Most of us learn at an early age about red, yellow, and blue. These are the colors of pigments. A pigment will absorb all colors of white light except that which is the reflected color we see. So, if white light is shown on a fully reflective pigment, we see white, whereas if the surface has green pigment, then we only see the green wavelengths of the white light.

The same principle also works for the light which transmits through a medium. This is how a lighting designer will place a color medium between a light source and the subject which will be lit in order to change the color of the light. Only the wavelengths of light not absorbed by the color medium are allowed to shine. Light has different primary colors than pigment. The three basic colors of light are red, green, and blue. For many video signals, this will be noted as RGB.

The combination of two primary colors yields a secondary color. So, again, as we learned when we were young, the combination of blue and yellow (primaries in pigment) will result in green. However, the means to get yellow (more of an amber, to be honest) in light is to combine red and green primaries. Between a primary and secondary is a tertiary color, such as yellow-green. Placing the color variations in a segmented pie shape can order them into what is known as a color wheel. This helps to visualize how colors work together, known as color wheel theory, which allows designers the ability to create appealing palettes to the way we observe color. This includes the ability to choose complementary colors, which are represented by colors that are opposite each other on the color wheel. Today, a color wheel may not just be divided like a pie chart, but have concentric rings representing various shades (hue + black), tints (hue + white), and saturation (hue + gray) of the primary, secondary, and tertiary hues (pure color).

BOX 4.2 THE COLOR WHEEL

The color wheel dates back to the mid- to late 1600s and is attributed to Sir Isaac Newton's work on the visible spectrum of light. Newton likened the division of color to that of music notation. He would begin with red at one end and continue through the visible colors. Like music, which repeats in octaves, the color wheel did as well. Purple, being a non-spectral combination of violet and red light, was thus placed next to red.

The color wheel was further refined by Albert Munsell who also wanted to bring order to the study of color. As a musician might be able to "hear" how a composition might sound by reading a sheet of music, his idea was to be able to "see" colors based on 3D attributes of hue, value, and chroma.

The use of color theories can help to make decisions which are logical and useful in how colors are used. Color harmony is visually appealing. It can help the viewer experience a sense of order and balance. Disharmony in a color palette can not only be chaotic (which could be sought after) but can visually bore the viewer to the point of disengagement. As the human mind is offered countless amounts of information constantly, only the parts that are engaging are readily perceived. We reject under-stimulating information. So harmony is the dynamic equilibrium between under-stimulation and extreme complexity.

In order to find color harmony, the designer should reference the color wheel. The simplest form is monochromatic, which by itself can elicit a certain feeling. A color scheme based on analogous colors, three colors which are side by side on the wheel, is one method of choosing a color scheme. Usually, one of the three colors will be dominant in this method. Another choice may be to choose complementary colors. These are two colors which are directly opposite on the color wheel. This creates the maximum contrast between the hues. For a slightly greater range of contrasting colors, a split-complementary scheme can be used. This will use three colors where the main color picked will have the

BASIC CONCEPTS 71

Figure 4.3 A simple optical illusion where the matching gray tones in the centers can appear as two different tones.

two adjacent colors to the complementary color. Another scheme using three colors is triadic, which uses three colors equidistant on the color wheel (such as the primaries), but should use one as the dominant color with the other two as accents. The most complex scheme is known as double-complementary, using four colors. These colors are each adjacent to a complementary color. Using four colors does result in greater difficulty in balancing them, but when they are harmonized in this scheme, the results can be spectacular. When looking for inspiration, it is generally a good idea to take a look at nature.

A color can appear different when it is in close proximity to another color. There are some great optical illusions which illustrate this (see Figure 4.3 simple and Figure 4.4 complex). This does not just apply to grayscale images as illustrated here, but to colors as well. This is known as simultaneous contrast (*The Law of Simultaneous Contrast of Colors* was published in 1839). All this says is that when two different colors come into direct contact, the similarities seem to decrease and dissimilarities seem to increase. The effect is at its greatest when colors are directly contrasting in hue. The eye can even be tricked into thinking like-sized objects are actually different in size, depending on how the colors react with each other. These illusions will appear different in print, direct view displays, and as projection.

Figure 4.4 A more complex illusion, illustrating the Bartleson-Breneman effect where the chess pieces are the exact same tone, but the varying tones in front and behind the pieces make it appear that the pieces are white and black.

PATTERN

The next element is pattern. A pattern is a visual arrangement of other elements in a repetitive fashion or some kind of sequence. They can produce interconnections and obvious directional movements. Patterns are not just limited to a decorative element such as a wallpaper design, but are found all over in nature. The subtlety of a pattern can be its strength, which is why they can be overlooked in the natural world. In the urban world, patterns are much more obvious, such as the design of a fence or the bricks in a wall. Also, pattern is not restricted to shapes, but can include colors as well. All of these can be used as a tool to grab a viewer's attention. Humans are built to recognize patterns as a method for memory. The pattern itself can thus provide a soothing psychological effect on the viewer.

The ordered arrangement can also represent a transformative ability which can be seen in contemporary artwork. For example, in *Day and Night* by M.C. Escher a checkerboard of farm fields (a recognizable pattern) transforms into another pattern, that of a flock of migrating birds. By using pattern in this way, visual excitement can occur. This may also elicit emotion or meaning, such as fear (or patriotism) in rows of marching soldiers or joy coming with a conveyor belt filled with treats. Or, as Andy Warhol intended, the repetition of soup cans identified the uniformity of advertising.

TEXTURE

Texture in video is simulated, for texture is something that we experience through the sense of touch. Visually, we see how light and shadow play across a surface and that can be replicated visually. However, through our experience of knowing what something looks and feels like, texture can create a physical response. The idea is often to create a trompe l'oeil (literally to "deceive the eye"), which is when we are able to create a subject to the point that its depicted form could be mistaken for a natural form. With other artistic mediums which are physical in nature (sculpture, painting, etc.), we can recognize a physical texture, even if the signs tell us not to touch. A texture can be abstract, being derived from an actual surface appearance but altered in some fashion to suit the artwork. The texture can mimic a natural texture or can be wholly invented through the imagination of the designer. An invented texture may be inspired by a Surrealist artist or may be used by the designer to represent something alien.

Textures may have patterns to them. A pattern does not require dimensionality; they can be simply artistic ornaments for the designer's purpose, whereas textures are there to represent how something may feel. A texture can be described as smooth, like glass, or rough, which would have a 3D disruption of what would have been smooth. The texture in the overall composition is seen as variations of light to dark, elevations, and recessions. If an area of the design requires more attention, enhancing the texture can transform it from being lifeless to exuberance with some slight shading changes. An example of how a visual representation of texture can evoke a sense of how the objects feel is illustrated in Figure 4.5.

Figure 4.5 The Larder by Anton Maria Vassallo.
Source: Image in public domain.

In order to give a design a sense of space, the designer may want to vary the quality of the texture. In order to make something seem more distant, the designer should mute the color variations (contrast) and slightly blur the texture, as evident on the right side of Figure 4.5. However, to make an object seem closer, the texture should be sharp (greater contrast in colors with more defined edges). This is the basis of a principle known as atmospheric perspective.

SPACE

Space is the measurable distance between points or images. Understanding how it was used in art throughout history can provide distinct stylistic choices in a design. How it is used or misused can have a dramatic effect on the composition of the design. This is something that can be even more of a challenge when taking the 2D design derived on a computer monitor and then projecting it on a 3D surface (the use of a UV map, a 2D model of a 3D object, can help). This is different from perspective, which is a graphic system of presentation to give the illusion of 3D objects, including creating spatial relationships on a 2D surface. More on this as we discuss the design principles.

As was discussed with texture, either sharp or diminishing detail of objects is a clear indication of spatial relationship. This can allow the designer to create a limited space, such as an interior setting, by limiting the depth of the design. The confinement can be any object which would restrict the view of a farther object, such as a wall or a hedgerow. In addition, a stylistic representation of many ancient art forms utilized shallow space in their art, such as in ancient Egypt. This may make the design appear flatter (which may be an intentional choice) than the use of deep or infinite space. The greater the depth of space, the more open the image appears.

Other means of defining space include their size, position, and how they overlap. We understand that two objects which would be equal in size if next to each other will appear as different sizes based on their distance to the viewer. This variation in scale can force the viewer to look at the 2D image as one of the third

Figure 4.6 *The Marketplace in Bergen op Zoom* attributed to Abel Grimmer.
Source: Image in public domain.

dimension. However, this illusion can be quickly diminished when compared to real objects which do not change in scale at the same rate (an actor or set piece moving up/down stage from the perspective of the audience).

As can be seen in Figure 4.6, many spatial principles are used. In the lower left, a person is in the opening to a cellar (restricted view). The detail of the cobblestone diminishes as perceived distance increases. The hills in the distance have muted colors, indicative of distance. Variations in size of people and objects imply relative distance to the viewer. If an image similar to this were to be used as a backdrop, the designer would want to appropriately size the image to closely match the height of the closest people to that of the performers, to avoid spoiling the illusion.

Though size may be a determining factor of spatial relationships if the objects are known to be of the same size, placement of different-sized objects can provide depth and importance in viewing relationship. A larger object generally has more prominence than a smaller one in our perception. Generally, this is due to our instinct of needing to deal with a closer object more than something farther away. A closer object may still be smaller than a distant object due to being dissimilar in original size. To give it prominence, the placement within the composition should be such that it gives the viewer the idea that it should take precedence as the closer object. When we have a perspective horizon, objects that are either higher on the upper portion of the perspective or lower on the bottom portion of the perspective will seem closer. In addition, objects which partially obstruct the completion of another object will appear as if they are in an overlapping plane. An overlapped image will give a much more powerful indication of spatial distance than any other spatial indicator. Again, back in Figure 4.6, the cathedral is obstructed by other buildings, which also helps to determine its size beyond just spatial orientation.

Space utilizes the other elements described previously. Line, which implies continued direction of movement, provides the viewer with that direction (as used in perspective). Lines are tools to help clarify spatial dimensions of solid shapes, especially when in relation to one another. Shapes, as described, are objects which occupy space, and will vary in size and varying means of distortion to represent spatial dimension of two dimensions to represent three. Color designates space as a means of lighter colors being assumed to be in the foreground with a virtual light source illuminating them more than those at a greater distance. The order of the value of the hue should be consistent as is linear perspective. In addition, a greater contrast in colors will give the viewer the perspective of an object being closer than one with more muted contrast.

In addition to the principles of space and the relationships of objects (providing volume to an image) is the concept of negative space. This is the space between objects. It is a means for minimalistic design, maintaining a composition with limited elements. It can be devoid of design elements (solid color) or implied absence from muting of background elements. It can be used either to separate or connect elements in the design, creating an emphasis on the different graphic elements of the composition through larger spaces or relationships with the use of narrower spaces. How this space is used can help the design convey different meanings. As with size and placement of an object, the size and placement of negative space can promote a visual hierarchy (active negative spacing) and promote visual recognition of the elements (passive negative spacing). Passive spacing is particularly necessary when dealing with text, where ill-defined spacing between letters/words/lines can make the text difficult to read, while greater use of spacing can give better comprehension. Particularly clever uses of negative space can also create new shapes, giving additional meaning to the overall composition.

> **BOX 4.3 BEST PRACTICE: MATCHING SHADOWS**
>
> There is a special note in regard to light and the virtual subject that the designer should heed. We know that a solid object will block light from continuing, creating a shadow. In the natural world, there is a singular source for light, the sun. This creates shadow in a predictable manner. On stage, multiple sources of light are used which can alter how shadows are cast as well as their appearance (not being as sharp or dark) and number. Not working with the lighting designer could mean that the design of shadows in a projected image could be opposite that of those cast by the artists and objects on stage. This is just one of many pitfalls of creating space with projection.

Spatial perception is a mental conditioning of the observer using their experience of the environment which they have experienced. It is good to remember that our brains have to interpret the information collected by our eyes with gaps of information filled in by experience. So how we perceive space may have a lot to do with our own experiences; thus, how space is represented in a composition can be viewed differently by the audience. The concepts of how space is perceived are by our stereoscopic vision (having two forward-facing eyes which each get slightly different views of the world at the same time provides us with the ability to see in 3D space) and kinesthetic vision (experiencing space through the movement of the eye, where the unconscious mind attempts to organize a 2D surface into separate parts to be seen as a whole, creating a spatial illusion).

COMPOSITIONAL PRINCIPLES

Using these basic elements, the compositional principles of harmony, variety, balance, rhythm, dominance, proportion, perspective, and movement will be used. Depending on who teaches these compositional principles, there could be fewer or a greater number, or known by different names. These are less set than the basic elements, but just as important.

HARMONY

Harmony in music is understood to be multiple notes played at the same time by different instruments (including voice) which have a relationship, building a cohesive whole of the musical composition. They must sound agreeable and are often at various pitches. Harmony stems from Latin and Greek (*harmonia*), referencing a joining or agreement of a concord of sounds. The same can be said of visual compositions. Visual unity is about elements working together, such as color schemes and the use of complementary styles. In addition, harmony does not have to be in individual scenes, but in the production as a whole – how transitions are made between scenes or where and how visual elements are displayed can either be harmonious (working together) or disharmonious (elements which work against the joining of parts). The level of harmony used in the design can have an impact on how the audience feels about their experience. The greater the level of harmony, the more comfort and at ease the audience will feel.

In Figure 4.7, we can see a sense of harmony in the style in which the image was created. Through a pattern of almost entirely vertical lines, an image is created. The consistency in compositional style grants harmony due to the rhythm and flow. In addition, the color pallet is limited; the simplicity adds to the coherence.

Figure 4.7 *Wuxtry!* by Albert Abramovitz.
Source: Image in public domain.

As an extension of harmony, unity adds order to make content feel like a coherent whole. Visual unity is how the elements work together, through schemes and styles. Often this will use repetition of a color or element to achieve consistency. Also, there is conceptual unity, which combines elements through blending form and function in a natural way. This form of unity is for the observer's convenience, providing visual clues which are familiar from prior experience. Unity in the composition should be able to achieve three points: that the element has a good reason to be present, that it works well with other elements, and that the concept is clearly communicated.

BALANCE

Similar to harmony is balance. Symmetrical arrangement, where elements are equal from one side of the composition vertically or horizontally to the other, can add a sense of calm. Symmetry can be reflectional (where the two halves are exact mirror images), translational (where the same shape or element is repeated on both sides of the design, as seen in Figure 4.2), or radial (where elements radiate from a central point). Content which is out of balance can create a sense of unease. However, something that is not symmetrical can still be balanced. An asymmetrical arrangement can still provide a feeling of balance, but creates a more dynamic feeling in the observer. This could be as simple as one large object on one side with multiple smaller objects on the opposing side. Physically, this could be represented by an equal weight of lead and cotton, where they have the same mass but are distributed differently.

VARIETY

It is said that variety is the spice of life. This is true of design as well. While the projection design should always be in support of the production, blending seamlessly with the rest of the production and not distracting from it, there is the risk of it becoming too lifeless. The design elements should still have something visually interesting for the audience. If not, the design runs the risk of dragging the rest of the performance down with it. Variety keeps the audience engaged. The easiest method is utilizing juxtaposition or contrast. This can happen between scenes (going from a dark and dreary interior to a brightly lit exterior) or elements of one area being harsh and angled to soft and curved in another.

This does not need to avoid harmony as each scene can still be harmonious, even if there is considerable discord between them.

Refer back to Figure 4.1. The contrast of being on the shadow side of the bridge in contrast to the bright daytime image breaks up the monotony. In addition, the highlighted area under the bridge showcases some activity over the stationary individuals in the shadow on the opposite side of the bridge. In Figure 4.5, the contrast of vibrancy of the animals with the dull nature of artificial objects helps to showcase the variation in the still-life image.

RHYTHM

As with music, a visual image can also have rhythm. It may have some underlying tempo which leads the audience to view it at a certain pace. This may be overall shapes or colors, through repetition. Rhythm can give a static image a feeling of action or movement. A random rhythm will have repeating elements without any sense of regular intervals as opposed to a repetitive pattern of elements which are similarly sized and spread out over predictable intervals. The rhythm may be less noticeable, such as those expressed in a flowing rhythm which is found in natural patterns such as the stripes of a zebra. More dynamic patterns expressed in a progressive rhythm will have a gradual change in the elements, such as gradients in color. Rhythm encourages the viewers to move their eyes across the design, following lines and forms. In Figure 4.7 there is rhythm in the spacing of the lines. The syncopated variations of color give a sense of life.

FOCUS

Contrary to rhythm is a sense of dominance or focus. This is where the audience is ultimately drawn to a point of emphasis or what could be interpreted as the most important "thing." This gives the audience a sense that they have purpose in what they are observing and not being lost in what they should look at. This may provide the director with a chance to hide clues in the media designer's images, by drawing focus away from a visual element which may be necessary later. This effect is achieved through the manipulation of elements (line weight, color, greater size of a shape, etc.) to make a specific part stand out. As we have alluded to previously, in Figure 4.1 the artist utilized a couple of elements to draw focus. Line and color specifically draw the eye to the ladies doing their wash. There are other action elements which are less noticed as they are in darker areas and without particular elements which draw the eye. In addition, the other brightly lit areas of the image are essentially lacking definitive focal points, such as the cloudy sky.

PROPORTION

One design principle which can be particularly challenging, especially when dealing with video backgrounds, is proportion. This is the relationship between two or more elements, which may present a challenge when comparing to the live element. In order to remain "proportionate," the elements require coordination between the elements. As larger elements can take emphasis in a composition, it is good to make sure that size is still correct in regard to surrounding elements, so as not to seem out of place, since size can also indicate proximity of an object. When the proportions of elements are too drastically out of the norm, they can be distracting. This does not preclude the use of proportion to purposefully distract, such as to be used for a production of *Alice in Wonderland* or a character who may be hallucinating. With objects where size is arbitrary, a more liberal use of proportion can be used for accentuating it without the cause of making the design seem unbalanced.

As previously stated, if a backdrop image were to be created similar to Figure 4.6, foreground elements would dictate the size that the image is presented to provide a proportional backdrop. There are individuals who come to the base of the image, giving the spatial relevance which needs to be matched by any performer who would be near the image. In comparison, Figure 4.1 does not have the same sense of proximity to the viewer. This could allow the designer to play with the image size depending on need (size of projection area, for instance).

PERSPECTIVE

Perspective as a spatial representation will have a horizon line (the eye-level viewpoint of the observer) and objects will work towards a vanishing point based on that horizon. The lower the horizon line is created, the greater the perceived height of the viewer and vice versa. How the image looks will depend on whether the perspective uses one, two, or three vanishing points. Most often, we consider a single vanishing point. This is a single point on the horizon line in relation to which all structures will appear to reduce in size as they approach it. This directs the observer to feel as if they are in a single location with a singular view. In order to expand where the viewer will look, two points are created on the horizon line. Generally, the points will be beyond the space of the composition.

If there is the desire to create the impression of great height, a three-point perspective can be used. For this, the horizon line will be close to either the top or bottom of the image. There will be two points on the horizon line, as with a two-point system, but then a third vertical vanishing point will be opposite the horizon line. This is known as a "frog's-eye view" when the horizon line is low, giving the impression of objects of great height to the observer or conversely a "bird's-eye view" where the horizon line is near the top of the image, giving the perspective of looking down on the scene. The closer together the vanishing points are, the more distorted the image will become (similar to a "fish-eye" lens on a camera, though less rounded).

Linear perspective has been used for centuries, but on its own it may not allow realism in the image being created. In order to add flexibility and realism (depicting more natural appearances), artists have found ways to improve these techniques. This may include adding additional vanishing points, creating depth of the objects. The linear perspective, especially with great depth of space, can better emphasize a single point for the observer, but may have the illusion greatly destroyed when viewed by observers in a large audience with varying viewpoints horizontally and vertically.

MOVEMENT

Finally, the principle which may seem to fit most with projection design is the idea of movement. This is not necessarily meaning that the objects have actual movement, but a sense of movement. The arrangement of objects or position of figures can represent this. This may include the use of lines or adjustment of shapes and colors to create focal points which encourage the viewer to see the composition in a directed manner. This can also be a perceived movement with the change, such as a series of lights that look like they are moving when in sequence every third light turns on, with the trailing two off, and then sequentially turns off while the next light turns on. There is no actual movement, but it directs the eye to follow the brighter area in the direction of the pattern. The idea that there is movement can provide the audience with a visual design which can be more lifelike without the need to have actual movement which could inadvertently draw the eye away from the performance.

Of course, the projection designer will often use actual movement (not still images). Once displayed, movement is the most forgiving (for quality of the image) as we do not focus our attention at any given point. As other physiological elements impact the viewer (brighter object, closer objects, sizable objects, etc.) then movement must also be considered. Instinctually, we respond to movement more than to any other visual stimuli. This means that a great amount of motion in imagery can take focus away from other parts of the performance. The designer needs to keep this in mind as the intent is on how the visuals will impact the overall performance.

As can be seen in Figure 4.8, this still image implies movement. Waves in the water suggest movement. The men in the foreground, pulling their boat ashore, are in poses which suggest that they are actively working and not at rest. This is in contrast to the ships anchored offshore, which hint at them being stationary. However, the observer will have their focus pulled in two directions due to contrast differences. The centrally located moon illuminates the clouds and surrounding scene. Meanwhile, the bright

Figure 4.8 *Moonlight* by Claude-Joseph Vernet.
Source: Image in public domain.

campfire draws the viewer towards the lower left. Yet, with this still being an unmoving image, it would less likely distract the audience to any physical movement happening on stage. So if the designer is to use an actual moving image, slow subtle movements are less distracting from the live performance.

The public domain images seen in this chapter are all converted to grayscale for use in this text. Each one is an amazing piece of art and should be appreciated in color. These can be found on the websites of the museums hosting them.

- *The Ponte Salario* (www.nga.gov/collection/art-object-page.41665.html)
- *Wreath of Laurel, Palm, and Juniper with a Scroll inscribed Virtutem Forma Decorat* (www.nga.gov/collection/art-object-page.50725.html)
- *The Larder* (www.nga.gov/collection/art-object-page.46190.html)

- *The Marketplace in Bergen op Zoom* (www.nga.gov/collection/art-object-page.57418.html)
- *Wuxtry!* (www.metmuseum.org/art/collection/search/374079)
- *Moonlight* (www.nga.gov/collection/art-object-page.214332.html)

BOX 4.4 SUMMARY

- Projection design includes the basic <u>elements</u> of line, shape (form), color, pattern, texture, and space.
- It also includes the <u>principles</u> of composition including harmony, variety, balance, rhythm, focus, proportion, perspective, and movement.
- The use of the elements and principles of artistic design should give the designer the means of discussing concepts with the director as well as providing means of development of content.
- An understanding of these core concepts can help to fix elements that seem "off."

CONTENT

What is content? In the current era of social media, content is the short videos and posts created to be consumed by the masses as a method of communicating ideas and opinions. It is something that is required to be constantly created or renewed to drive traffic to a website. In the realm of live entertainment, it is something to be expressed through an audio or visual technical medium. It is one of the three components necessary for the experience, the other two being the space where it is consumed and the technology used to display it. For the modern projection designer, it will be some form of digital file which contains a still image or motion picture, or even some form of manipulative elements and specialized coding for interactive designs.

Without content, the projection system is useless, as that is a tool to deliver the content to the audience. The system is the blank canvas, the paint, and the brushes for the artist to present their masterpiece, but it is not a masterpiece until it is used. The content is the ideas we have been exploring thus far. They are the ideas, which may be difficult to acquire, but are essential to drive the story.

The designer needs to take the inspiration of the planning phases of the mood board/storyboard/action chart, combined with the principles and elements of composition, in order to develop the

DOI: 10.4324/9781003107019-5

material which fulfills one or more of the fundamentals of design. This is the content. Whether the designer is utilizing pre-rendered (already created) work or creating the work from scratch, there will be a number of elements required to observe in order to ensure that it will effectively communicate the desire of the director. As the content is acquired, the designer will refine the ideas, meaning that the first composition will rarely be the final product.

Each design should be uniquely inspired. This means that the designer may not want to approach the development of content the same each time. Again, it should be emphasized that the designer should have the end goal in mind when making decisions about the rest of the design. Certain things will be essential to one design while practically unnecessary to another. No matter what these are, they should give meaning to and support the story.

Content is going to come in the form of many different components. These may be simple images or graphics or more complex video elements. Its layout can help establish meaning to a production, as can the color or themes chosen. The type of imagery that the designer chooses to use can have unique psychological impacts on the viewer that may or may not be intentional. It must be delivered on schedule. A timeline for the creation, review, and delivery of the content must be adhered to, ensuring all media is ready and tested prior to installation.

CONTENT DESIGN FUNCTIONS

While all visual media utilizes the design fundamentals discussed in the last chapter, with projection design there is also the method of how content functions within the overall production. These are not hard and fast rules in the same way that the elements and principles of composition can be. The functions are more of a guide as to the means that a designer might employ to support the production. The following six functions are meant to advise the designer on how the content will be used.

INFORMATIVE

The first function is purely for information. An informative design function may be required by a script to identify the time or place of a subsequent scene. Those of us who have seen a production of *Les Misérables* will know that scenes are defined with places and dates to understand the progression of the story. This information is projected for the audience so that they do not have confusion of where and when the action is happening, releasing the playwright/lyricist from requiring dialogue to provide that information. Even if not required by the script, the director could make a request to help make an identification so as to move the story forward with ease, preventing the audience from pondering on that information.

When providing information, legibility of text is highly important. Thus, decisions about the size and type of font used as well as the contrast of the text to the surface that it is projected on require an understanding of how the audience will be able to see and understand the information. High contrast between the characters and background will give the audience the best opportunity to be able to read the information provided. This is recognizably more difficult when the text falls beyond simple titles, as mentioned above, and gets lengthier, such as a quote.

Other factors will also come into play regarding the legibility of the text. The difficulty in stating certain specifics is that there are many variables at play in how visual designs will be represented. All of these will impact how the observer will interpret the information provided. However, there are a number of guidelines which can help the designer in creating the best possible scenario for the text to be legible (if that is the desire).

The size of each character will be dictated by the distance of the furthest viewer. This is opposite to the desired pixel size, which is determined by the closest viewer. The size of pixels and their density can have a negative impact on the near observer in not being able to distinguish details if too large or greatly spaced. As an observer gets farther away from the display, pixels appear to blend together, allowing for greater size of pixels and space between them (in the case of direct view displays, more on this in Chapter 6). For an optimal character height, a general rule is to divide the viewing distance (in inches) by 172, which would give you the height of the character in inches. The smallest character should be no more than the viewing distance (in inches) divided by 215. If the distance is unknown, the text should be at least 2–4 percent of the height of the display. Any characters created smaller than these recommended guidelines can make it increasingly difficult for the observer to be able to accurately determine the information. One thing to note is that text which is created as an image and then scaled larger (within the projected image, not just an increase of the projected image) may develop visual artifacts (mistakes) as opposed to actual text which is changed in font size as one is a raster image, while the other is a vector image.

BOX 5.1 RASTER AND VECTOR IMAGES

So, what's the difference between a raster and a vector image? These are two means of creating a graphic representation in a digital image. Each will have its own pros and cons and may need to be considered separately when creating images. Most design elements for live entertainment will contain raster graphics, but may occasionally require vector graphics, especially if there is a need to change the scale of objects.

A raster image is composed of pixels (picture elements, a physical point in a picture which is the smallest addressable element representable) in a specific resolution (number of pixels horizontally and vertically). This collection of pixels is known as a bitmap or bit array. The image is able to refresh independently of the complexity of the image and has a wide range of colors. The greater quality of the image will mean that it will be a much larger file size. Various algorithms will compress the amount of information to reduce the file sizes. Scaling down an image is relatively easy as details are taken out to reduce the pixel count, but scaling up will create challenges and will often make the image pixelated (jagged) or blurry as the computer attempts to interpret means of filling in spaces without information. Common file types are .jpg, .bmp, .tif, .gif, and .png.

A vector image, on the other hand, is composed of mathematical paths (sequential commands or mathematical statements). This results in smaller file sizes, not requiring algorithms to reduce the file size. Larger images may flicker when the display refreshes as this method requires a redrawing of the entire image. Curves,

> polygons, and other primitive shapes are drawn as continuous and smooth lines as opposed to just being an approximation. Changing the size of a vector does not introduce anomalies in the image. Vector images may require specialized software to open, and especially to edit. Vector images are important if there is a need for an element to be sized differently, such as a logo. Common file types are .svg, .emf, .eps, .ai, and .dxf.
>
> Digital displays use raster images, so a vector image will need to be converted. Once converted, the vector (now a raster) will have all the properties of the raster. So, if the image needs to be scaled larger, the designer will need to set the size prior to conversion so as not to introduce poor qualities into the image. The opposite is not true. Changing a raster image to a vector will not really give the positive properties of the vector. The quality will be poor and the image size is often reduced.

In addition, the informative function is often required for supertitles. This is the captioning method for translation for productions like a foreign language opera, often displayed above the performance area. As a constantly changing text, which should not distract from the performance, legibility and understanding need to be rapid. In order to do so, the style of the letters used should be simple. For instance, using all capital letters has been found as a method to quickly discern text (which is why street signs use all capitals). The challenge for the designer will be less artistic, but more in how it is displayed to be read. This will come more to play later as the methods of display are discussed. For more detailed information regarding recommendations for text height as compared to image size, the farthest and closest viewer and their viewing angles, the Display Image Size for 2D Content in Audiovisual Systems is a standard (ANSI/IFOCOMM v202.01:2016) which can be referenced.

When text is less relevant to advancing the story, all concerns about viewer distance, stylized lettering, and contrast can take a back seat to the artistic choices of the designer. Typography can offer a lot of information on its own. The perception of the viewer can also allow them to pick up on a lot of what is desired to be seen, even if it is not actually there. "Can yo rea   thi?" It is very possible that the audience can fill in the missing letters and answer "yes" to that question. This falls back to the way we see and how our minds will fill in the gaps.

In addition to text, image magnification (IMAG, pronounced "eye-mag") is another means of providing information to the audience. While this is most often seen in concerts or conference presentations as a means of using a camera to capture the performer/presenter and projecting the close-up image on a screen, it does not need to be that simple and can have much greater creative looks. The purpose is still to give the audience the ability to see something on screen in detail that they cannot see otherwise. This does not need to be strictly a camera feed, but can easily have added effects or a focal point other than a performer.

SCENIC

The next function is one of the most likely to be utilized in a production, which is to create a scenic element. Scenic backdrops have been a staple of theatrical productions, allowing for a space-saving opportunity to provide a visual representation of a specific location in order to develop the acting space. These scenic elements can be costly as they require large pieces of scenic cloth and long periods of time for talented artists to paint them. Each scenic location requires a unique drop, compounding the time and financial commitment. Buildings which contain a fly loft (space above the stage for scenery to be lifted out of view of the audience) have a greater opportunity to use drops as they have the ability to show multiple locations through the use of multiple drops. The time and space required to paint one drop, let alone many, often put this beyond the reach of many productions; hence the need for rental companies to provide drops of popular scenic elements. This leads to producers and directors thinking that replacing a painted drop with projected images will be simpler, and probably cheaper. While there are financial benefits in the long-term investment, the initial purchase cost may be staggering if not fully prepared. There are additional benefits to replacing a traditional backdrop, such as adding motion elements, creating a more cinematic feel, especially when combined with automated scenery.

As stated, a director may be interested in replacing a painted drop or multiple drops with one that is projected. This can raise new challenges and considerable benefits. The initial outlay of a purchased projection system may be beyond the realm of many productions, requiring rental solutions (some of the companies that provide digital backgrounds also rent out the projection equipment) which can still be an extraordinary expense. One challenge is that scenic backdrops are rarely in the same aspect ratio (the ratio of height and width) as that of the video projector, requiring either the concession that the projection will be a different size than desired or a more complex video system will be needed to achieve the desired results. In addition, a traditional backdrop is lit by the lighting designer, and any additional light, particularly from follow spots, may spoil the effect a bit. This does not necessarily mean that projection should be avoided; but with any projected backdrop, light from any source other than the projector can have a significant impact on how the image is seen. Having a direct-view backdrop may be a better choice (more on this in subsequent chapters).

Projected scenery is not restricted to the backdrop, but can cover all visible surfaces. It is also not limited to a single static look. A traditional drop may resemble a forest, but a projected backdrop can add life to that forest by making it appear to have a breeze move the leaves and branches. Clouds can change in the sky instead of being cold and unmoving. But if the scenic design never required a drop, projection can still provide wonderful backgrounds, such as looking through windows to see "outside," or simple projections on the flats to provide other scenic elements.

This function of projection design can easily have the same pitfalls as traditional painted backdrops. Altered perspective is a method to force the audience to believe that they are seeing something different from what is actually being represented. However, this can be quickly dismissed by the audience when perspective is not matched. As soon as the performer approaches a projected image which is mismatched in scale, there will be a jarring change to what the audience observed prior to that interaction. The difference between a painted backdrop and a virtual backdrop is that the latter has the ability to actively change. With prior planning, as a performer approaches the background, the scene can change to keep the perspective alive.

Beyond replicating a traditional scenic viewpoint, the point of view can help the audience feel as if they are seeing the action from a different perspective, especially considering that the projection is not relegated to just the backdrop. This could potentially give the director the ability to show just how perilous a position a character may be in by giving a "bird's-eye view" of a drop below.

A virtual scenic environment can allow the director to create a more cinematic feel to the production. While automated scenery brings life to a show, preventing pauses that were traditionally required for the stage crew to set the next scene, automation enhanced with video elements can create effects that used to only be possible in cinematic productions. However, the video image can suffer the same fate as the traditional backdrop in the matter of scale, especially with forced perspective. The benefit of video is that it has the ability to be adjusted given time and ability during technical rehearsals.

EMOTIVE

As a means to drive the story forward, the director will likely want to elicit an emotional response from the audience. Simple scenic elements may not be enough to support the dramatic elements being presented by the performers. Images do not need to strictly represent some tangible element of the real world. Additional content which is not directly tied to the action can enhance the emotional reaction of the audience. In a function similar to a movie score, where different types of music lead the audience to a variety of emotions, visual media can also have this power. The director may be helping the audience to understand the emotions of the character or leading the audience toward a general mood of the story.

While this function may seem melodramatic, it can be a useful tool. As the lighting designer may utilize specific color schemes to help to elicit an emotional response, so can other designers. Colors may be culture-specific, being understood differently depending on the audience. For instance, as described in *If It's Purple, Someone's Gonna Die* by Patti Bellatoni, films with a dominant red influence have themes and characters that are powerful, lusty, defiant, anxious, angry, or romantic. These emotional themes follow cultural responses from Western culture and how people of European descent understand the meanings of color. However, red in Chinese culture represents luck, happiness, joy, vitality, and fertility, so is used for celebration. It is also a color used by brides to ward off the auspices of evil. As such, the designer needs to understand how audiences understand color as this could drive the emotional response in the wrong direction.

Color is not the only tool to guide emotions. Symbolism common to a given audience is very useful. Cultural symbols can be used to this end, but this can have some differences depending on who the audience will be, potentially limiting cross-cultural relevance. These images may be a direct representation (or slightly distorted), such as a photograph or motion picture, or they can be more abstract. An image of a flag could be a scenic element if presented in a realistic manner in the way it would normally be seen. However, if that flag is represented in a non-scenic manner, then it may take on a stronger role in how the audience perceives the narrative. Consider if the flag is on fire or is cartoonish with the colors bleeding from it. This type of media, with the purpose of affecting the audience, will not directly be in existence in the world of the character. However, it may have a direct relationship to the emotion of the character.

SPECIAL EFFECTS

There may be many reasons for a director to call upon the projection designer to add to the story through the magical use of special effects. This can be very subtle or overtly impactful, creating a moment of amazement. Historically, this could be argued to be the primary purpose of theatrical projection. From the early days of the phantasmagoria to the iterations of Pepper's ghost, projection is an ideal means to create something which may or may not exist in order to fool the audience. In modern times, the director may turn to projection for reasons of safety (i.e., projected fire in place of real fire), budget, or the needed experience to create a practical effect. The projected effect could also accentuate a practical effect to make it appear more than it actually is, such as adding to the volume of an object. This type of design will likely be exclusive to projected effects and not direct view (such as LED walls) for the designer.

In 2012, at the Coachella music festival, the late rapper Tupac Shakur was virtually resurrected to perform "in person" with the actual performer, Snoop Dogg, stunning not only the audience present, but millions afterward as they watched recordings on the internet. While this was not a new technique, the use of modern materials and quality content made it extremely popular. It completely overshadowed a similar performance of Madonna and the Gorillaz (a band who uses avatars in place of the actual performers) at the 2006 Grammy Awards. The techniques used to create this type of effect require immense planning and cooperation with other departments, with the risk of failure much greater than for other effects.

While techniques of Pepper's ghost (the above effect) are impressive and garner a lot of attention, other projected effects may be so subtle that they are virtually unnoticed by the audience. Projectors are light sources. While this may seem to be obvious, they can function as a lighting instrument. A projected image, which can be completely white, may be one of the brightest single lights on stage. If something on stage needs to be highlighted, an image can be manipulated to have black in areas not intended to be lit, allowing for complex objects to be illuminated that would be impossible for standard lighting instruments with framing shutters.

Somewhere between the extremely subtle and the highly impressive is where the bulk of special effect projection will fit in. It can provide a fire in a fireplace where actual flame cannot be used. It can give performers "superpowers" such as the appearance of lightning coming from their fingertips. Such effects often require the performer to be immediately behind a semi-transparent material such as a sharkstooth scrim. On the other hand, projection is not limited to the set and props, but can be used directly on the performer. The performer's costume may be altered with projected effects, or even virtual makeup has been used.

For much of the special effect function of digital media, the designer is adding to reality. This is known as augmented reality (AR) or extended reality (XR). Unlike virtual reality (VR), which is an entirely personal experience, where the audience members are individuals with their own unique perspective with a single video display for an all-digital environment, AR and XR can be more shared experiences. AR is a technique where digital content is overlaid upon real-world elements. This often requires an individual viewing device similar to that used in VR, but will also require a camera to bring in the real environment. While much of the public around the world may have been introduced to AR in 2016 with the mobile app Pokémon Go, festival attendees at Coachella in 2018 had their music experience enhanced with their own app allowing them to see items from a bowl of spaghetti (Eminem's performance) to a Coachella-branded space shuttle. XR should not require an individual intermediary device for the audience, but is shared with all in techniques like Pepper's ghost. Not to put too fine a point on it, this function will often take more preparation for its effectiveness than other functions of video.

TEXTURAL

Whereas scenic images can provide a recognizable element to a surface, a textural image will be for the quality of aesthetics (not the texture design element). Similar to the emotive function, this adds characteristic features to the production which are not always realistic or driving specific elements of a storyline. While this function may elicit an emotional response, its purpose is separate from the emotive function. This design function is more often used outside of specifically theatrical productions and is more often associated with live music backgrounds, or potentially behind presentation materials of a corporate event. However, they can blend nicely in a dance number.

The imagery for a textural design will be abstract, not meaning to represent anything in particular with what is happening in the performance space. These visuals may elicit some added excitement and are there to give "life" to what would otherwise be "boring." Textural elements may or may not run the risk of being a distraction from the performance. This will depend on the amount of contrasting colors, the amount of movement, and the luminance in comparison to the main subject, as well as many other elements. The image size will often be very large, similar to other scenic backgrounds. If used with dance or other performances where the focus should still be on the performers, there is the risk that this can be an upstaging visual, which can be the point for many music festivals where the focus is on the overall experience, not just the performers.

INTERACTIVE

Finally, the director may want the projection design to be interactive. Interactive projection may be the most specialized function as it will often require more elements to the physical design than just the content (media source) and display. It is often a function which enhances prior design functions. This will likely require additional precision not always required of the other functions. Interactive design will require considerable preparatory work which could be outside the realm of a typical production schedule. Due to this, the designer may need to offer alternatives to the director to fulfill the story ideas. If it is in the realm of possibility, this can be spectacular.

What is commonly imagined for interactive projection is the manipulation of content by the performer or some other outside entity. Oftentimes a video element appears to be reacting to a performer, but it is actually just well choreographed, similar to a dancer to music. True interactive video will be able to react to the performer or some other catalyst, regardless of the timing of the choreography. This will require additional cameras for motion capture and a variety of sensors, as well as different means of handling content. Interaction may simply be a trigger for content starting or may be a method to alter it. Some content is even rendered live, not having substance that can play without interactivity.

The designer can avoid some complexity in interactivity by having an operator trigger a video sequence instead of having a sensor or some other means of the performer triggering the same sequence. But, interactivity is more than just a means of starting and stopping video playback. It often includes techniques which allow the content to have the sense of being a part of the living canvas. Additionally, a non-traditional projection surface can include objects that move throughout the performance space organically. This, of course, requires some means of following the moving object and computational ability to rapidly alter the projected image to match the orientation of the object in regard to the position of the projector.

One of the most stunning uses of interactive projection is when a character is developed. This may be a digital puppet, where sensors are determining the movement of a live actor and manipulating a wireframe skeleton of a digital avatar seen by the audience. As opposed to pre-recorded content, this allows the digital character to react as does any other actor, playing on the audience in a much more personal manner.

Interactive projection will require more technical knowledge and will often require the designer to closely work with other design teams for it all to work. Many effects can be created with off-the-shelf components, not always requiring specialty items. Generally, the most complex portion of the design is the media control getting the appropriate data in order to use it for manipulation. While the results can be spectacular, it will take serious consideration to know whether it is worth the time and effort required or whether another solution may be better.

COLOR IN CONTENT

The original image presented in Figure 5.1, the inspiration for the musical *Sunday in the Park with George*, shows how composition is

Figure 5.1 *A Sunday on La Grande Jatte* by Georges Seurat.
Source: Image in public domain.

of utmost importance to the main character. The artist Georges Seurat was influenced by various studies on color theory. The creation of his "system," which we call chromoluminarism or pointillism, makes an impressive study for the digital designer. This style of painting matches the technology of direct-view displays, where adjacent individual dots of color are mixed by the viewer's eyes. Seurat's belief was that this would ultimately provide a more vibrant appearance.

BOX 5.2 *A SUNDAY ON LA GRANDE JATTE*

To fully appreciate *A Sunday on La Grande Jatte*, the reader is encouraged to view the full-color image provided by the Art Institute of Chicago where it resides: www.artic.edu/artworks/27992/a-sunday-on-la-grande-jatte-1884

As previously discussed, color has different meanings to different cultures. A designer may choose to use a particular color scheme in order to influence the audience or may choose to use a natural flow of colors to fit a realistic setting, without emphasizing any one color or group of colors. Understanding how color will be displayed may influence the choices of color (what is seen on a computer monitor can be changed by the type of projector and the surface it is being projected upon).

Emotions affected by color have been studied at great lengths, especially by marketing firms who hope that their choices will impact the consumer in regard to the product to be purchased. That same psychological impact can be used for emphasis for the rest of the content, especially when it may not be descriptive of the objects represented. Simply increasing the value range (the contrast of light to dark hues) can give life or direction to a particular color scheme. While certain color schemes will be cultural, others will be more universal due to shared experience.

When colors are very similar to natural objects that may be common to everyday experiences, they may reinforce ideas along those lines. Imagine the color of a lemon. Lemons don't vary in color much in nature. When an object shares that same color, it may have an impact on how that object is perceived. Consider three drawings of a sword – one which is realistic in a steel design, one that is black, and one that is lemon yellow. Might the perception of that sword be different to the observer, even though the object has not changed apart from its color? Might the one that is black even be considered wicked, maybe more so than the realistic sword?

Light and bright colors tend to make us feel more joyful or uplifted. Warm colors which are leaning more towards the red spectrum are generally stimulating. Cool colors, those trending towards the blue portion of the spectrum, seem to be calming and tend towards vast expanses. Dark colors tend to be more somber and depressing. Even subtle variations of these color schemes have been used to impact visitors in trauma centers to sports locker rooms.

How we feel about an object can be heavily influenced by the colors of surrounding objects. In a grocery store, raw meats will generally be in an area that has a lot of white, giving the idea that it is clean (to ease concerns of contamination). In addition, contrasting leafy greens will be placed around the meat which will emphasize the reds, giving the impression that the meat is fresher. Thus, the designer should consider in their composition how surrounding elements may impact the message of the content, including elements not projected.

Too few colors may have an impact as might too many colors, especially if we are not looking at natural color schemes. This goes along with not only what is projected, but the surrounding sets and costumes and how they are lit. After all, the projection designer needs to constantly remember that the choices made by one discipline can have a direct impact on others.

TEXT IN CONTENT

Text may be a critical component of the design, even when it is not being used for informative purposes. The legibility of the text may or may not be of concern to the designer. The style of the font and how text is presented can often be as descriptive as the text itself. Consider when a group chooses to make an invitation to a Halloween party, they may choose to use a font which appears to look like blood dripping. On the other hand, that would probably be a poor choice to use for a child's birthday party.

If the text is meant to be read and there is the desire to use stylized text, then the designer needs to take into account the size recommendations of the informative function. In addition, consider how much time the audience will have to read the text, especially in ensuring that it will not be distracting from other dialogue. The more difficult it is to interpret (size of letters, stylization, or length of text), the longer it will need to be displayed.

Text itself can become a work of art, either by stylizing the lettering or by creatively spacing the text. There are several forms of art which use writing to create photomosaics, each having their own unique style. These forms of text art include ASCII art, word clouds, calligrams, ambigrams, and concrete poetry. Each form emphasizes a visual format that exceeds the text itself.

Utilizing ASCII characters as picture elements to create an object has been practiced from the time of typewriters (an example of a butterfly was created in 1898). The simplest form of this is used to create emoticons, simple graphical representation creating facial glyphs to represent emotions. ASCII art can give a feel of the early days of computers where it was necessary to use this style, for printers could not print other graphic elements and were limited to the 95 printable characters. This form of imagery can be easily created through any number of programs where a photograph or graphic is transformed into ASCII art. The images can also be animated (early animated ASCII art started in 1970) either by changing characters within the image or by a sequence of completed images. Again, there are modern programs that can create this effect quickly and easily from footage the designer may already have.

With the text itself in a supportive role, a classic form of word art is concrete poetry (a modern term for this form of text arrangement, used to enhance the meaning of a poem). Shaped poetry has been known since ancient Greece. It also has roots in some religious texts of Judaism, Christianity, and Islam. A modern approach to this art, where the presentation is the primary purpose, is the calligram, which arranges the text into a thematically related shape. In a typewritten image, it resembles ASCII art in that creative spacing of the letters creates a silhouette, and upon closer examination it is revealed that there is prose and not just elements used for various shading, though words may be broken apart for effect. In hand-written form, a calligram can be more free-flowing into a greater representation of the visual object.

Word clouds, sometimes known as tag clouds, are a form of weighting a list of textual data in a visual design. The size of each word in the shape represents the relative importance in comparison to the other words due to frequency of use, significance of meaning, or categorical quantity. These are then arranged in some form of pattern for visual effect. The larger size, with the indication of greater importance, will definitely resonate with the audience as it will be most visually impactful to the greatest number of audience members. The less important, hence smaller text size, will be viewed by fewer audience members. Of course, like the calligram, the importance of the words may be less important than the shape they create. Many text clouds used purely as reference will be an amorphous shape (thus a cloud), but the size of the words and their orientation can form discernible shapes.

The final form of word art to take note of is the ambigram. This is a calligraphic design which allows for multiple meanings depending on how it is viewed. This can be a much more difficult design challenge, but may make a strong visual impact. It can be very representative of dual nature, which may be thematically relevant in a design. There are several variations in this category. Graphic palindromes, which will describe the majority of ambigrams, have their origins as far back as the ancient Greeks. A simple palindrome will have the same meaning if the letters are read forward or backward, though it may be necessary to rearrange spaces between letters. Rotational ambigrams can either be natural or non-natural. The complexity of ambigrams may or may not benefit a projected design.

COPYRIGHT

This text should not be taken as legal advice on copyright permissions or the exact method on obtaining rights to use the work of others. There may be any number of misconceptions about what is and is not legal to use which float around on internet forums, especially in regard to fair use (a legal doctrine that permits the unlicensed use of copyright-protected works in certain circumstances) for educational purposes. This is a battle which the designer may need to fight in order to protect themselves or their work. Laws in most developed countries are put in place in order to protect intellectual properties (ideational property, such as patents, methods, and processes, as well as literary and artistic works). This is to ensure that the rights of the owner (individual or corporation) are safe from unlawful infringement of that property.

For the designer, the most common protection which will need to be considered is the copyright designation. This is the legal right granted to an author, a composer, a playwright, a publisher, or a distributor to the exclusive production or distribution of a literary, musical, dramatic, or other artistic work. The extent of these legal rights varies by jurisdiction and the designer should consider where their work will be used and the size and type of audience prior to requesting rights, as that may impact the cost and approval to use.

Content which is not specifically wholly created by the designer should be considered to be under copyright protection, as all works in the United States created after March 1, 1989 did not require a copyright notice to protect the work. In other words, copyright protection is implied for all creative work unless it is specifically published with a statement that it is freely available to use. This means that even if the media is found in some publicly available forum, such as in a library or the Internet it should not be considered freely available for use. To prevent unauthorized use, some artists will put an identifying graphic over the top of the finished product, known as a watermark. This makes the work undesirable to use in that state. However, this section is more about the designer knowing when creative elements by others can be used and not the protection of their own intellectual property from others.

For many artists, their craft has been honed over many years and their work is their sole source of income. Therefore, even though a production may want to avoid costs wherever they can, even a not-for-profit company should take the proper channels of acquiring rights (be it paid or just credit given) for all content not created by them. There are options where content can be freely used and may be the best option for those productions on a tight budget.

Intellectual property can have the copyright expire after a specific amount of time. This is generally past the date by which the creator or beneficiaries would be receiving compensation. This generally will make the work quite old. The dates on which an item becomes part of the public domain (a non-legal term used to describe a work not protected under copyright law) can vary depending on when and where it was produced, how it was distributed, and when the creator passed away, potentially making challenging research for the designer who wants to use a particular work.

It should be important to note that, according to the United States Copyright Office, there is no amount of changes that can be made to another's work which would allow for a claim of new copyright. That being said, a derivative of a pre-existing work will in some way incorporate the original work but revise it into a new work. For a projection designer, this allows something like a sculpture to be used and turned into a digital representation. There is still a form of creation.

> **BOX 5.3 ROYALTY-FREE CONTENT**
>
> Stock content from a royalty-free provider is an option that can be quite beneficial to a designer. Often, content is provided in different resolutions and formats, making it beneficial on the occasion when a production is remounted and requires a slightly different version of the same content. The designer may not be required to purchase the rights to use the content again, just download the required content. The downside of online stock content is that a website could ultimately shut down, preventing future downloads of purchased content. In addition, since the royalty-free content is used under contract, there may be challenges to find who owns the copyright for proper attribution or whether the content can legally be used for future productions.

ACQUISITION

Not every designer will have the ability to create digital content wholly on their own. They may not be able to hire a team to do it for them either. For this reason, there are a number of organizations which offer content or elements of content whose rights to use can be purchased. This is known as stock content.

Stock content is not the same as the prepackaged content (mainly digital backdrops) provided for popular shows, which was previously discussed. Instead, these files are elements to be used in whole or in part for a licensing fee. The content is not created to represent any particular work, but provides elements which can be used for a variety of purposes. A number of websites will offer this content, some producing all of the content, while others are host platforms for independent artists. Media provided will range from digital photography to computer-generated graphic elements, animation to video footage. Some video content will be created in a seamless loop while others are just a clip of action.

Fees involved with stock content can vary greatly. This will either be with royalties or as royalty-free. Media which is distributed under a royalty contract will likely be a complete work and will contract a fee to be paid for every use. Rehearsal use may be without cost, just paying per performance. More often the content will be royalty-free, meaning that there will not be a fee per use, but a fee upfront. This is not a purchase of content granting full rights; therefore, the designer cannot distribute the content to others. The specifics for use will be noted on the site from which the media is acquired.

Another means of acquiring content is to look for works provided to the public for use. An American non-profit organization known as Creative Commons has aided artists in providing works for others to use. Individual works may be available for others, as deemed by the creator, with the attribution requirements listed with the work. Some images in this text were discovered through sites associated with Creative Commons.

Regardless of how attribution or payments are required for stock content, this is a valuable tool for the projection designer. The time and cost of creating the visual elements by the designer or their team may be the deciding factor in choosing to use a stock element. For instance, a transitional element requiring a castle could be created as an animation, but if all other projected elements were photographic, then the designer may choose stock footage instead of traveling to a location to film it themselves as that may cost too much as well as take too much time.

Stock elements will rarely be something complete for the design and will require manipulation. If the designer does not have the skills to create or manipulate the content, they may require the assistance of others. These can be members of the team or the work can be hired out to third-party groups. It would be beneficial to have third-party groups not use stock material in order to keep rights management evident. Any deliverables from a third party should include associated attributions.

ASPECT RATIO AND RESOLUTION

While projection design has considerable flexibility in how it is seen and perceived, the technical elements will provide some constraints which need to be considered when creating the content. It is recommended to "think outside the rectangle," or not to feel constrained by the borders of a projection screen or extent of a video wall. While this may be necessary at times (using a practical display, such as a television required by the script), it should not constrain the creative process.

When creating content, it will ultimately be formatted to a given width-to-height ratio, the aspect ratio. There are a number of common aspect ratios for traditional delivery (movies, television, computer screens), which will be available in every editing program. The ratio will either be expressed in whole numbers, such as 16:9 for high-definition (HD) television standards, or fractions compared to one, with 1.778:1 being the same aspect ratio. Table 5.1 includes a selection of display resolutions along with their associated aspect ratios.

When content does not match the aspect ratio of the display, the two options available are to omit portions of the content or to omit portions of the display (more on this in Chapter 8). There is a third, less desirable option – to distort the content by stretching or compressing the content to match the aspect ratio. Omitting portions of the content was a regular option for when cinema was brought into home televisions, which had drastically different formats. More than just omitting portions of the image on the sides (allowing the full height to be filled), a process known as "pan and scan" was used to ensure that the relevant portions of the image were not omitted. In the same scenario, as consumers were able to afford larger televisions, and the audience wanted to see the full

Table 5.1 Standard display resolutions and their associated information

Aspect Ratio	Resolution	Designation	Naming convention
4:3 or 1.33:1	800 x 600	SVGA	Super VGA
	1024 x 768	XGA	eXtended Graphics Array
	1400 x 1050	SXGA+	Super eXtended Graphics Array Plus
	1600 x 1200	UXGA	Ultra eXtended Graphics Array
	2048 x 1536	QXGA	Quad eXtended Graphics Array
16:9 or 1.77:1	1360 x 768	WXGA	Wide eXtended Graphics Array
	1600 x 900	HD+	High Definition Plus
	1920 x 1080	FHD (2K)	Full High Definition
	3840 x 2160	4K UHD	4K Ultra High Definition
	7680 x 4320	8K UHD	8K Ultra High Definition
	1280 x 768	WXGA	Wide eXtended Graphics Array
16:10 or 1.6:1	1920 x 1200	WUXGA	Widescreen Ultra eXtended Graphics Array
	2048 x 1080	2K DCI	★Digital Cinema Initiatives 2K
~17:9 or 1.85:1	4096 x 2160	4K DCI	★Digital Cinema Initiatives 4K
~21:9 or 2.35:1	2560 x 1080	UWFHD	★Ultra Wide Full High Definition

theatrical release, portions of the screen were not used, presenting only black above and below the content, known as letterboxing.

Beyond consumer displays, professional projectors will also have set aspect ratios. The same principles can apply as above, where the content can be altered to fit the display, though many projectors will also offer a setting known as blanking. Blanking is a process by which the display will omit variable portions of the image vertically or horizontally, allowing it to be almost imperceptible that the projected image is different from the display surface.

The aspect ratio of content is determined by the resolution of the image – which is the number of picture elements (pixels) used to create the image. This is often matching the display where it will be used. For the Advanced Television Systems Committee (ATSC) standard for "full HD," the resolution will be set to 1,920 x 1,080 pixels. The list of standardized display resolutions, of varying aspect ratios, is quite long. Many of them are named acronyms by the Video Equipment Standards Organization (VESA). Some of these naming conventions are part of the common lexicon, even for consumers: HD = 1,280 x 720, while ultra-high definition (UHD) is 3,840 x 2,160 (also known as 4K), both of them having the aspect ratio of 16:9. Knowing these naming conventions is useful when comparing displays, as some manufacturers use them while others use the numerical representations. Also, more pixels within the same visual area (increased pixel density) leads to the image looking crisper.

Basic editing software will limit the designer to a few standardized resolutions when rendering the content. Advanced display techniques will generally require non-standard resolutions. Even when the design is simply to fill the cyclorama with a static image, the aspect ratio of what can be seen by the audience is rarely that of a standard display. Knowing how the content will be displayed will be important in determining the level of detail required in the design.

BOX 5.4 PREVISUALIZATION

Before too much effort is put into creating content, it is good to show the director and the rest of the designers the concept of the projection design. Previsualization (colloquially known as previs or previz) is a crucial process in the design process that involves creation of a digital representation of the projected content and its interaction with the physical environment. This technique allows the designers and other stakeholders to visualize, plan, and refine the projection setup before the actual installation and execution.

This process is known throughout a wide variety of creative processes, including filmmaking and video game design. Essentially, this is a technique of creating a visual media to explore the development process that portrays the end point of the designer's vision. This is fundamentally an advanced form of storyboarding, which may have already been used to advance initial ideas. Previs can be achieved through animated 3D-modeling software, visualization platforms, or even simple digital images, illustrating the content where it will be shown.

As the set designer will need to have plans off to the build team early on, the projection designer can start showcasing how the end design will look using the renderings of the set. This can give the director a better understanding of what to expect in the final result. In order for the director to make informed decisions, the previsualization should reflect how the final design will look as closely as possible, including any imperfections, so that expectations can be discussed prior to loading in the space. Several options may be shown to offer the director a means of adjusting expectations based on relative capabilities, such as brightness versus resolution.

Previs models will provide the basic visual framework in planning the flow and transitions of the projection content. This can help perfect the aesthetic of the overall design. The process will offer the designer a means to better discuss sightlines (both of the audience and the path of projected light) as well as positioning options. This can save a lot of time and money by focusing on what works and what needs more work. The process can be used internally, allowing the projection team to simulate interactive elements. This may help fine-tune triggers and responsiveness of the system. It may also identify potential issues in the physical setup, such as obstructions in the projection path.

FORMAT: CODECS AND CONTAINERS

All content will take a certain amount of memory as a digital file. The broader the range of colors and more pixels required the larger the file. In order to reduce the size of the file for storage and use, a system of compressing and decompressing the file is used. A codec is a computer program which encodes and decodes the data stream for this purpose. First off, as there are different types of media, including still images, audio, video, and general data files, each one will have a different means of dealing with file size. Thus, a codec for one type of media may not be able to handle another. There are many different codecs which all have different pros and cons as to the file size compared to the quality of replication.

Codecs are essential for media compression by reducing the size without significantly compromising the quality which enables smoother transmission. Some software is specifically designed for one step of the process, but not the other. When the data is encoded in a particular format (DivX, MPEG-4, H.265, etc.) then the counterpart decoder is required in order for that data to be displayed. As technology advances and file sizes increase, new codecs are required, with their own benefits and drawbacks.

All codecs can be divided into two basic categories: lossless and lossy. A lossless codec will compress the data in a manner that essentially retains all of the original information. This allows for near-perfect reconstruction of the original data when it is decoded. This is done by only eliminating redundant information. This is of particular importance for future video editing or live compositing. Meanwhile, a lossy codec will achieve higher compression ratios by discarding bits of data which have less significance. Often, this type of codec is based on the limitations of human perception for how the file is recreated. Lossy codecs are of particular importance in situations where storage is very limited or the bandwidth available is limited (such as streaming applications).

While the detail of how the process works is beyond the scope of this text, it is important to know that the two styles of compression are intra-frame and inter-frame. An intra-frame compression is a type of spatial compression. Each individual frame of video is compressed independently without taking the surrounding frames of video into consideration. This technique is particularly helpful in sections of video which are relatively static (frame to frame) or sections lacking a lot of detail. This technique could result in choppy playback in complicated or high-movement scenes. For this reason, this type is less often used with lossless codecs. In contrast, the inter-frame compression is a type of temporal compression. It takes advantage of the temporal redundancy between consecutive frames of video. This type of compression technique identifies and encodes only the differences, including motion vectors, between frames. It will reference previous encoded frames (key frame) and eliminate redundant information until the next fully encoded frame. This allows for higher compression ratios, particularly in videos with repetitive motion patterns. As a consequence, the decompression process requires access to the previous frames for accurate reconstruction of the frames. This dependency means that manipulation of playback may be less than desirable.

The way picture data is structured in a stream or file is the frame type. There are three types of frame, known as I-frame (a key frame), P-frame, and B-frame. An I-frame is an intra-frame, which consists of only the macroblocks of information within the frame itself. It can only use spatial redundancies (similarities between pixels of a single frame) in compression techniques. A P-frame is a predicted frame which allows compression using temporal prediction in addition to spatial prediction. This way change is predicted by previous frames as well as nearby information (or entirely skipped). This means that P-frames require I-frames at regular intervals since P-frames only refer to previously encoded pictures. Meanwhile, a B-frame, or bi-directional frame, can refer to frames that occur both before and after it. This form of frame type can be very efficient in reducing the size of the file while retaining video quality, but can be resource-heavy at both encoding and decoding the file.

The codec is only one part of the equation. A container is a bundle of the video data (codec) and metadata (data that describes other data), organizing it into a single package. This provides the file extension, such as MP4, AVI, MKV, MOV, and so on, that is seen and misunderstood as the file type. The metadata can include information on the type of codec used (so that the playback device knows how to decode the video), information about when the file was created, and language files (including subtitles). Consider a can of soup, where the can is the container, with a wrapper that gives information on what is included (list of ingredients), when it was created (or that can be surmised by the expiration date), but does not do anything to give you satisfaction (that's the point of the soup). Without the label on the can, it is impossible to determine the contents of the can. A playback device needs that data, just as it is necessary to know if you will have a dinner of soup or pet food.

Choosing the right format will depend greatly on how the video is to be used and how it will be played back. This means that the requirements of a specific system will determine the format. Certain software requires specific codecs. Knowing which system will be used will often determine the choice of codecs. The designer needs to know what program/system is going to be used prior to creating content; whereas having content that was formatted incorrectly, requiring transcoding (converting one format to another), may cause degradation. In addition, there are several additional settings used when rendering files which can have an impact on file size.

SPECIAL CONSIDERATIONS

When creating content, there are other parameters that can impact compression and the quality of the finished video. These include bitrate, color depth, frame rate, and the resolution/aspect ratio. Some of these choices will be made at the start of the content creation, while others will be made at the time the video is rendered. A rendered video is the end process where raw (and sometimes layered) footage, images, and graphics are converted into a playable format. This process encapsulates all of the frames of video (some may take varying extra time to process depending on the complexity), audio mixing (if included), applying compression, and exporting it to the finished file in the specified destination.

At the beginning of the creation process, considerations such as resolution and aspect ratio are chosen. If using a standard display, such as a single projector or monitor, then this will be determined by that display. However, if multiple projectors are being used to create a larger image, or if a video wall (made of multiple monitors in an array, or LED panels) is used, then the resolution and aspect ratio may be non-standard. The content is best created to match this display methodology, which could be beyond the scope of simple editing programs. Creating content in a resolution or aspect ratio that does not match the display may end up with the content looking distorted or imperfect in some way. However, higher resolutions do require more processing power and storage space, so some additional components may need to be considered in order to balance the file.

Color depth and type is the next consideration to be established at the beginning of content creation. The two basic variations are cyan, magenta, yellow, and black (CMYK) and red, green, and blue (RGB). The former is subtractive mixing (beginning with white and working towards black) used primarily in the printing process. The latter is additive mixing (starting with black and working toward white) and is the primary method of creating video images. The third variation is the addition of an alpha channel. This is an added color channel which is stored as an absence of color in the areas where this is designated. This creates transparency where there is no color, allowing background images to be seen in those areas, in a layered video. Each color system will have variations on how many colors can be created, which is known as bit depth. The more colors available to be created, the greater the file size will be.

Frame rate, the number of frames displayed per second (fps), is a consideration both at the beginning and end of the content creation process. Cinema, along with some television broadcasts and streaming content, has traditionally been displayed in 24fps (some modern movies at 48fps, to mixed reviews); while video has a variety of frame rates (25, 30, 50, 60, 120, 240) along with some conversion rates (29.97, 59.94). In life, fast-moving images look out of focus, an optical effect known as motion blur. When we no longer see motion blur in fast frame rates, our mind perceives this as illogical and causes disorientation and nausea, accounting for why some high-frame-rate cinema may not be accepted by some audiences. FPS for video is often based on power standards available in the region. So in areas where 60Hz is the standard, video frame-rate standards are best created to match, where other areas of the world have 50Hz standard (Hz is the designation for hertz, the number of oscillations of electricity in an alternating current per second). If frame rate is too slow, we may perceive a flicker between images. Obviously, the higher the frame rate, the greater the amount of image information required to be stored, which can have a dramatic impact on the file size.

Beyond how fast the video is intended to be shown is how fast the data can be processed to show the video. Bitrate is the data which can be processed per second, denoted as kilobits per second (Kbps) or megabits per second (Mbps). File size and picture quality will be determined by the number of bits available. There are two means of processing the bitrate: constant bitrate (CBR) or variable bitrate (VBR). In CBR, the processing of video is constant, regardless of the complexity of each frame. Meanwhile, VBR encoding will vary the output of bits over time, depending on the complexity of creating each frame.

The process of rendering the completed video can be quite time-consuming and taxing on the computer hardware. As previously stated, there are a number of choices that will impact this, including the choice of codec. Due to this, a good practice for the designer will be making some changes in rendering early on in the development phase to ensure the content will meet expectations prior to fully rendering the files. If content will be managed by a media server that will manipulate the content, rendering a short clip and importing it into the server for use will provide information on how well it will work (from the import process to manipulation in the server) in case changes will be required in the rendering process. In addition, rendering a full clip in minimized settings (lower resolution, color space, low bitrate, etc.) may allow the designer to illustrate to other members of the production team how the end result may look. By taking some additional steps early in the process, the designer can save a lot in time and resources.

AUDIO – EMBEDDED OR NOT

One additional consideration in the content creation process is the choice of including audio in the video file or not. This is a consideration that will depend heavily on the production itself, how it is created and how it will be staffed for presentation. There are several reasons for and against embedding audio into the video file. As with most other considerations, the inclusion of audio may come down to experience and design.

The first consideration should be how the audio will be de-embedded. While the consumer does not usually need to worry about this (most televisions have speakers, or can pass signals to external speakers), a production for an audience will likely have a separate audio system from the video system. While most video content will be played on a computer, the audio connection may not be suitable for theatrical presentation. On some occasions, a media player with separate audio outputs could be used. This may or may not provide quality audio signals.

Benefits to having the audio embedded are coordination of audio and video playback. This is most impactful when the disjoining of audio from video is evident in clips such as a person talking (lip sync issue) or a sound effect (sound of a bullet being fired). Having audio and video processed together can solve timing issues as there is no operator error stemming from multiple operators needing to start the content together. However, as will be seen in a later chapter, show control can solve this issue.

One of the biggest challenges of including audio with the video file is that there may be two completely separate design teams. This could require the audio or projection designer to finish their work in order for the other designer to complete theirs. Due to the length of rendering times, a last-minute change could potentially become unattainable. In addition, any live effects desired by the projection designer could be hindered by the addition of an audio file. While an audio file might be used to develop the video content, most often the inclusion of that file in the final video rendering will not be preferred.

BOX 5.5 SUMMARY

- There are six main functions of projection design: informative, scenic, emotive, for special effects, textural, and interactive.
- Copyright laws vary by region and it is very important that the designer be familiar with the specifics of the area where the production will be shown.
- Content can be acquired from a variety of sources for use in a production. How it can be used and attribution required for use will depend on the source of acquisition.
- Aspect ratio is a relation to the resolution of an image, which may or may not match the display device or surface.
- Content is created using a codec and container, which allows for storage and transport of the files.
- Which format is chosen is most often determined by the playback system.

6

EQUIPMENT

A vital part of any design is the process of displaying the content. There are three critical components for a digital display: the source, the method of distribution, and the display device. More complex designs will include a network of devices and have expanded methods of control. While the method of display should be considered during the initial design process, how the content is delivered to the display should also be considered when developing the content. There are many variables that can impact the quality of the presentation.

The cost of various components of a system can vary greatly. Understanding the equipment types can help the designer and the projection design team to specify components that should fit within most any budget. What the audience is able to perceive should always be kept in mind, as "the best" equipment is often unnecessary due to the presentation variables. Due to the constraints of the venue, the designer may need to get creative with these three components to realize the design.

DISPLAY

The culmination of the design process will be how the content is ultimately displayed to the audience. There are two primary means of display, direct-view and projected. Direct-view displays

DOI: 10.4324/9781003107019-6

are commonly LED walls, comprised of multiple LED panels connected together to form a larger display, and are the complete display. Conversely, the video projector will require something to project the light upon or through for the audience to see. There are a lot more variables to projection than there are with direct-view displays. There are many advantages to using a video wall over projection, though it also has certain limitations only afforded to projected images.

One commonality among displays is that they have physical limitations on the image they can produce (actual number of pixels and color), but the signal processing may allow for the ability to show many variations, sometimes seemingly beyond those physical limitations. During connection, displays and the source will perform an action known as a "handshake" where certain metadata is shared, including the limitations of the display. The metadata format defined by VESA which provide this information has been known as Extended Display Identification Data (EDID). EDID includes data on the manufacturer and product type, the serial number, chromaticity, resolution and timings, luminance, and pixel mapping information.

As a means of keeping up with more advanced display technologies, and preventing the standard being encumbered by legacy architecture, VESA has developed a new standard known as DisplayID. This standard properly and efficiently communicates with advanced display capabilities. This now allows for dynamic refresh rates, extends field sizes for the support of high pixel counts, and provides support of high dynamic range (HDR, an improved aspect of video for greater range of color volume, luminance, and overall bit depth). As EDID/DisplayID are shared between the source and display, a poor connection can interfere, causing either no data to be transmitted (a blank image) or for the display/source to default to the lowest resolution available. Due to this, the management of this data is a crucial step and should have preparation prior to installation.

The video design team will have a great need to understand the capabilities of the display to match the content. This may require choosing the specific display before creating content, thus setting parameters for how the content will be created, or carefully selecting a display which will accurately provide the image intended. Whichever method of design is required, a basic overview of the two primary means of display should provide the designer with the means of discovering the direction of display technology to use.

VIDEO WALL

In the generic sense, any direct-view display comprising multiple display elements with distributive image processing can be considered a video wall. To make a distinction in the types of direct-view displays, a video wall excludes LED walls. So, what is left to make a video wall? The most common device is simply a collection of video monitors. Outside of entertainment, this is regularly seen in advertising as part of a digital signage solution or in television news studios replacing a green screen.

As video monitors are a commodity item, the cost may be relatively low. The challenge for the designer may come in dividing the content to be displayed across the array of displays. In addition, unless specifically made to be part of a video wall, video monitors lack the construction to be physically put together and a custom external frame may need to be constructed to support the array of monitors.

Monitors which are specifically designed to work as part of a video wall will have internal processing to properly divide a video signal into smaller components, allowing the whole signal to be viewed over the expanse of the video wall. Generally, these will be intended to be part of a digital signage installation and will have a dramatic price increase for each monitor as compared to consumer models. In addition, these will likely have additional individual image processing, allowing the end user to match the colors and brightness in order to balance the look of the entire display. These commercial units are built for longevity; something a designer should keep in mind if the production is planned on running more than a few weeks.

When using consumer model monitors (or televisions), additional equipment will be required to distribute the signal among the displays. Individual control will likely not be possible, but if all monitors were purchased together (same make and model) and have aged equally, then this may be an imperceptible difference. Another difference from a commercial model and consumer model is the size of the bezel, the outer "frame" of the monitor which does not play a part in the display. On a consumer model, which is intended to be viewed by itself, the bezel is not much of a consideration beyond aesthetics. However, when multiple monitors are combined into a video wall, the minimized (narrow) bezel of a commercial display causes minimal breaks visually in the overall image.

Processing the signal for a video wall can come in a variety of methods. A standard resolution image can be sent to select digital signage monitors, which can provide basic conversion of the image into smaller portions to be displayed in a set configuration (usually a matrix of 2 x 2 or 3 x 2 monitors). This will maintain the aspect ratio of the initial video file. For live entertainment, video wall control requires a processor, which can either be a dedicated hardware-based controller or software-based (requiring a computer with appropriate video cards). These will be discussed later in this chapter as part of the distribution components of a video system.

A video wall will most often be a single large display, but it does not need to be. The designer can choose to separate the monitors, leaving considerable space between individual displays. In addition, the displays may be mounted in landscape (horizontal orientation), portrait (vertical orientation), or on a diagonal. The complexity of processing does increase as the spacing and orientation is divested from a cohesive, singular image, but the results can be quite striking. In addition, when a designer runs into challenges (cost, power, or a plethora of other concerns), the flexibility of splitting the image may provide solutions. Even when the image is segmented, the audience will fill in the missing parts as long as enough of the image can be seen.

A video wall made of monitors can be very beneficial if the audience is close to the display. Video monitors are generally intended to be viewed from a relatively close distance, so the pixels within each display are tightly packed. This can result in a highly realistic image with an audience member being within arm's reach. The downside is that video monitors have some sort of surface (glass or acrylic) which can reflect other lights. In addition to the bezel breaking up the overall conformity of the image, the designer may choose to forgo this type of display except under specific design choices such as a picture window.

LED WALL

A specific form of video wall made entirely of LED panels is more suited to live entertainment. The LED wall has a great number of variables which the designer must consider when choosing to use this form of display. One noticeable difference between a video wall using monitors and an LED wall is that LED panels connect with one another virtually seamlessly, having no need for a bezel. However, there are a few other construction differences to note.

Unlike a video monitor, LED panels cannot process video signals on their own, but rely on an external processor. This processor will decode and distribute the video signal as established during setup. The complexity of the processor may only convert the signal or it may have the ability to control the "brains" of the LED panel, making adjustments to the settings in order to have them match one another. In addition, these more powerful processors can provide for non-traditional mounting options and the required distribution of signal to account for that. As an ultimately scalable display, multiple processors may be required to drive content over the entire surface, especially as the display may literally have millions of pixels.

Viewing distance for LED walls is generally greater than video walls made of video monitors. With LED panels, the viewer is looking at the light source per pixel. These pixels are clusters of multiple colored LEDs which use additive mixing to build towards white. The minimum viewing distance (how close the observer can be without being able to discern individual pixels) is a factor of the size of the individual pixel and how much space there is between them (pixel pitch measures the distance from the center of one pixel to its neighbor). The size of pixels and their density can have a negative impact for the near observer in not being able to distinguish details. As an observer gets farther away from the display, pixels appear to blend together, allowing for greater size of pixels and space between them.

As each pixel is able to be turned completely off, LED walls have superior contrast to other displays. The space between pixels is also made of light-absorbing plastics, reducing the amount of impact ambient light will have on the display. For displays which may be used outside, the protective mesh surrounding the pixels may also incorporate ambient-light-rejecting structures. These can impact viewing angles, but generally are acceptable for entertainment purposes.

Manufacturers of LED panels also make the mounting hardware. This includes rigging bars so that the wall can be flown, ground support options, and some are made for permanent mounting to a wall. The designer may need to consider the mounting options, especially for long-running productions, in how the panels can be maintained or repaired. Manufacturers understand that, in modern environments, a strictly flat display is not always desired. Due to that, most LED walls will be flexible in the mounting options, allowing for a curved display, either concave, convex, or a combination, allowing for a wave.

Although the most common LED wall will be a solid structure, there are numerous variations offering the ability to see through the display. This can be for visual effect or of a practical nature, especially for air flow in outdoor applications or for acoustic transparency (so as not to interfere with audio quality passing through). When using outdoors, the designer will likely need additional eyes on the design in the form of a competent person up to and including a structural engineer due to the forces that weather can impose upon the rigging, even with blow-through panels.

LED walls can also be made of a virtually transparent substrate, minimizing its visibility when not in use. Even with transparent displays which have little to no structure between pixels (imagine a gauze-like web of framing), the structure still needs support and the LEDs require power and data to them. This makes a structure that creates a visual reference as to where the tiles are, even when the signal is turned off. In order to make the power and data virtually invisible, manufacturers may be required to eliminate the scalability of the display. However, these displays are still large enough for most theatrical applications.

While LED panels are designed to lock together, there will still be vertical limitations designed by the manufacturer on how many panels can be suspended. Likely, most designers will never hit that limit, but even if that limit is not reached, the weight of the combined panels and rigging accessories can add up quickly. This may prevent the designer from using a display as large as they would prefer simply due to physical limitations of the venue or due to the cost of rigging support (riggers, motors, set-up time, etc.).

Another limiting factor is the amount of power required to drive a video wall. In the consumer market, LED technology is recognized as a power-efficient technology, especially in comparison to incandescent light sources. However, the LED panels still consume a lot of power. The manufacturer will provide maximum and average power consumption, which will give a good idea of how much will be required. The power consumption of most LED walls will actually be considerably less than posted if being used indoors. The listed power consumption is for panels running at full brightness; most productions will likely only need to have the wall at no more than 20 percent brightness. The brighter the colors, the greater the power need (a white background uses the most energy). The amount of power required is another factor where the projection designer and lighting designer (ultimately, the master electrician) will need to work closely together.

VIDEO PROJECTOR

The display technology which is widely accessible to media designers is the wide variety of video projectors. There are distinctions in how these projectors deliver the image between the light sources to the image processing, which will be more of a concern to seasoned designers who may find that they prefer one to another. These differences are more subtle than the variations in video walls. However, there are some variations which will impact the quality of the design, including the choice of lens, the brightness and contrast, the connectivity, and how the projector is controlled (the latter two may be of concern for video walls as well).

One of the most often-asked questions is "how bright a projector do I need?" If the goal was to only show a movie in a controlled situation like a cinema, then the guidelines set by the Society of Motion Picture and Television Engineers (SMPTE) state that a 12–22 foot diagonal screen should be illuminated at 16 fL (foot-lamberts) for a standard (not 3D presentation) film. However, in live entertainment, there are numerous variables which go into getting this answer, which will rely on a balance between science and experience. Too dark an image can not only prevent the audience from seeing the content, but dimming the rest of the performance could cause drowsiness as well. So, a lot will depend on how well the designer understands how people see and perceive what they see in order to make the decision over which projector will be chosen. Depending on the size of the projection team, this could be passed on to a team member who specializes in the technology to make these decisions.

BOX 6.1 ILLUMINATION DETERMINATION FORMULAS

There are three useful formulas (each a derivative of the others) which will assist in determining illumination.

Projector lumens / screen area * screen gain = foot-lamberts
fL * area / gain = lumens
lumens * gain / fL = area

How much light a projector can produce is measured in lumens. A lumen is a standard unit of luminous flux, or the power of visible light (that to which the human eye is sensitive) emitted by a source. The way this is measured has some variation; hence, a projector manufacturer will state the luminosity in a couple of ways. The two most common standards are American National Standards Institute (ANSI) lumens and International Organization for Standardization (ISO) lumens. A "standard" is kind of like a formula to describe the best way to do something, as agreed upon

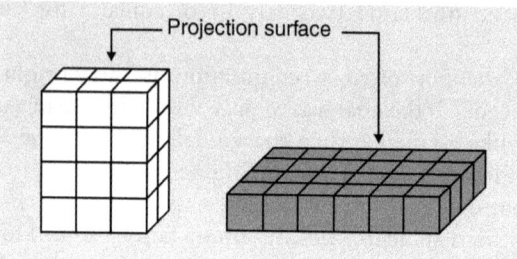

Figure 6.1 Block representation of light dispersion: the blocks represent the amount of light on a projection surface; due to the inverse square law, as the light is spread to four times the size, the image is a quarter as bright.

by experts in a particular subject matter. These two organizations have strict definitions of how the brightness is measured and under what circumstances. ISO and ANSI lumens are measured using a light meter at a variety of points of the projected image. This takes into account brightness uniformity and color accuracy. As the conditions required prior to measurement differ (environment and calibration), the resultant measurements will also be different. Quality manufacturers following these procedures will provide accurate readings which will help the designer choose a projector, but this is just part of the overall problem. If lumens are stated without the designation of either of the standards organizations, then the measurement will not provide much useful information and is likely a greatly bloated number.

A good starting point is to estimate the actual size of the projection (including any area blanked) and look to fill that with a minimum of 70 lumens per square foot. This is only a rough starting point as there are many factors that play into what is needed and which may cause the image to be less than expected. Some of these variables include the "truthfulness" of the manufacturer's stated luminance, the contrast ratio (including the amount of ambient light), the age and condition of the light source (and internal light path), the content, and the projection surface.

Even if the projector manufacturer listed the output of the projector, and was truthful in measurements, that would be only part of the dilemma. The second, and also regularly misrepresented, measurement is the contrast ratio. Unlike an LED wall, projectors have a constant light source with variations in imaging delivery (reflective or transmissive). This means that there will be some amount of light which is emitted, even when no color information is present for that pixel. The Audiovisual and Integrated Experience Association (AVIXA), an international professional organization, has developed a standard adopted by ANSI for measuring contrast ratio in projectors which more closely represents actual usage. A series of 16 rectangles in a checkerboard of black and white is displayed, measuring the light reflected at the center of each rectangle. The averages of the whites and blacks are compared to represent the ratio as compared to one. Under this standard, a ratio of 400:1 up to 2,000:1 could be expected. Some manufacturers will measure a "dynamic" contrast ratio where either the full white screen is measured to fully off or a reduction of the output in order to achieve a greater ratio. These numbers will be immensely different from those measured by the ANSI method. This information can still be of use to the designer in understanding how bright video black might be, in order to determine whether some other light blocking might be needed during blackouts.

The variation of contrast stems from the fact that not all light emanating from the projector is used to create the image. How well the manufacturer is able to control internal "light leak" (loss of light due to optics, optical coatings, the lens, etc.) will have an impact on how well the projector will produce black. Some imaging techniques, such as digital light processing (DLP), will have a better contrast than others, such as liquid crystal display (LCD).

Understanding contrast is crucial to determining the quality of the projected image. The manufacturers who exaggerate their published information know full well that the quality of an image and the audience's visual acuity will improve with greater contrast ratio. Looking for a contrast ratio in the hundreds (e.g., 800:1) instead of hundreds of thousands (e.g., 100,000:1) is good practice and will help the designer to understand how the image will look with content. The contrast ratio of the actual displayed image is often immensely different from what is stated by the projector, regardless of the means of measuring by the manufacturer. The "room" contrast ratio is a result of the projected image as well as ambient light (light other than the intended projected light on the projection surface). Even a small amount of unwanted light can reduce the perceived blacks in an image.

Most theaters will be designed in a fashion to minimize unwanted light. Many surfaces will be painted or stained black with the addition of light-absorbing fabrics hung throughout. But not all performance spaces are designed to control light this well. This means that the projector itself can be providing unwanted light after the image reflects off the intended surface and subsequently reflects off additional surfaces of the room, returning light to the projection surface. One aspect that can have an immense impact, but is not a light source by itself, is the use of atmospherics. Haze and theatrical smoke will fill the air with particulates, which allow beams of light to be seen (often a lighting design choice), dispersing some of the light uncontrollably.

BOX 6.2 EFFECT OF AMBIENT LIGHT ON CONTRAST RATIO

Consider a projector with 500:1 contrast ratio. This means that the whites are 500 times brighter than the blacks. Thus, if it measures at 2,500 lumens for the white image, the black image will be just 5 lumens. If the room has a slight glow, casting just 10 lumens' worth of light on the projection surface, this will add a total of 10 lumens to both the blacks and the whites (total of 15 and 2,510 lumens respectively, or a room contrast of 167.3:1).

Rarely advertised, but equally important is color contrast. The difference between white and black will give a good impression of the range of contrast in color as well. However, the imaging technology can impact this immensely. Higher-end projectors will give the projectionist the ability to make some fine adjustments on color production, which can be extremely beneficial in actualizing the design.

> Perceived contrast is a result of the human eye adjusting for brightness. This means that small changes are not particularly noticeable as the perceived change is not linear. So doubling of the contrast does not mean that we perceive a two-fold increase in the difference between black and white. However, it can change our perception on how bright the projector is over one with a lower contrast ratio.
>
> In addition, perceived contrast is subject to the Bartleson-Breneman effect. This is an optical illusion where the perceived foreground will appear lighter as the background becomes darker and vice versa. The more that black reaches an absolute black, the better the quality of the image. Due to this effect, the area surrounding the projected image (including the border of a projector screen) will impact what contrast is perceived. Examples of this effect are shown in the images back in Chapter 4.

While contrast and lumen output are two measured statistics which the designer can use to understand what projector to choose, there are several more variables that can come into play. While "get the brightest projector you can afford" may be common advice given on the Internet, this could be very wasteful in the overall production budget. It also does not account for the longevity of the projector (different technology and manufacturing processes can alter the quality of the image over time). Nor would this take into account potential physiological effects on the audience (too bright an image can cause eye strain).

Advice for how bright a projector should be must take into consideration not only the space where the projection will take place, but also the content. The standard for cinema takes into consideration a wide variety of content from bright animation to dark horror movies. This can come into play for the theatrical designer as well, knowing that there will be a wide variety of content. However, if it is known that all content will be bright colors with low contrast, the designer may be able to get away with a projector that ultimately is only producing 50 lumens per square foot (or maybe even less). However, if the content relies on high contrast, especially in the darker areas of the image, a much brighter projector may be required to maintain the desired contrast.

PROJECTOR LENS

Most commodity projectors, those that are intended for home or office use, will come with a standard lens that cannot be changed by the end user. This can either be a prime (also known as fixed) focus or variable focus (zoom) lens. Projectors which are intended for use in larger spaces include the ability to use different lenses. Sometimes the best choice in choosing a projector to purchase will be the availability of rental lenses, as purchase of a lens can be quite expensive and lenses are not always compatible with different models of projectors. If the designer cannot change the lens, then the only consideration is whether or not adjustment of the projection ratio is required and whether remote adjustment is needed.

A projection lens will have a number of designators to take into account. The two most important will be the throw ratio (compares the size of the image due to the distance from the surface) and the f-stop. Those familiar with photography will understand that the f-stop value relates to the size of the aperture, which determines the amount of light allowed into the camera (as the pupil does for the eye). Consider the f-stop for a projector lens as the reverse, aiding in the determination of how much light escapes from the lens.

Then f-stop for projectors is not a true reversal from cameras. For one, the depth of field that a photographer needs to pay attention to does not compare equally to that of a projector (primarily a projected image is on a flat plane). So, does this mean that the designer needs to just look for the lowest possible f-stop for the greatest possible aperture, allowing the most light to be emitted? This is mostly yes, but of course comes with some caveats.

The aperture of the lens plays an important part in how bright the projector is, but it also plays into controlling contrast. It is a delicate balance the manufacturers play in order to develop the best image. The designer will not have control of choosing two lenses of the same focal length with different f-stops. Instead, this is a matter of understanding how much light might change based on which lens is chosen, especially as there may be some overlap in throw ratios. Zoom lenses will have a range of f-stop values which depends on where in the zoom range the image is chosen. If not listed, then the value is indicative of the mid-range of the zoom. Meanwhile, a fixed focus lens will have a single value, as they also have a single throw distance. The designer will rely on the specifications provided by the manufacturer. If looking for a quick reference, one f-stop will decrease brightness by about 50 percent.

BOX 6.3 CALCULATING F-STOP STEPS

An f-stop step is not going to appear as normally increasing in whole numbers. A full f-stop is a mathematical calculation based on the aperture value (a proportional value of the opening which allows light through the lens). This is figured by taking the square root of 2.0 raised to the power of the aperture value. So a full stop is an increase of one aperture value. If the aperture value is 1.0, then the f-stop is 1.4. By increasing the aperture to 2.0, the f-stop is also 2.0.

The lens aperture conversely can affect the image sharpness. So while projections do not have a depth of field in the same way that photography will, the tolerance of focus increases with longer focal lengths (images projected over longer distances). This is known as depth of focus, which is the projection kin to depth of field. In technical terms, the longer focal length aligns the beam of light closer to parallel, known as collimated light. This is particularly useful when projecting on any surface other than a flat plane from a position perfectly perpendicular to that surface. Thus, the designer needs to understand that the longer the focal length the higher the f-stop.

The lens throw ratio represents a distance ratio from the projector to the screen where the image can be focused. This means that if a lens had a 1:1 ratio, the width of the image is the same as the distance the projector is placed from the surface. Where that measurement is taken can vary from manufacturer to manufacturer, so for precise measurements, it is good to use a projector calculator provided on the manufacturer's website, if available. For a wider image and a projector being close to the surface, a short throw lens is used (roughly designated as a ratio of less than 1:1). If the projector is required to be a great distance away, a long throw lens is used (roughly designated as a ratio greater than 4:1).

There are trade-offs to having the projector close or at a distance. Some of this depends on what is to be focused on (flat or 3D surface), whether shadows are a concern, and positions available for mounting the projector. It is important to know that ANSI lumens are measured under specific criteria for the size of the image, which can actually be different when changing out lenses due to the f-stop. It is good to remember that the amount of light per square area is going to be considerably lower with a larger-size image. To visualize this, refer back to Figure 6.1. Additional light is lost in a long throw lens, though the depth of focus is greater. A short throw lens will lose the least amount of light, but needs to almost exclusively be directly in alignment with the flat surface to avoid visual distortion.

The choice of lens should also include considerations of geometry and distortion. As the primary lens design is intended for a flat field (projection screen), the focal length is designed to have all areas of the imaging medium (all elements within the lens) focus within the same plane. Zoom lenses are particularly difficult to manufacture to these standards and inferior quality may exhibit aberrations such as "pincushioning" (where a rectangular image will look as if the sides are being pulled to the center, with the corners appearing streched) or uneven focus at the extreme ends of the zoom. Meanwhile, a short throw lens may have the opposite effect, known as barrel distortion. In addition, an effect known as spherical distortion, caused by uneven focus of the light rays as they pass through the lens, can create areas out of focus within the image, preventing the entire image from ever being in complete focus. How the projected image is used and the type of content to be seen will determine if this is acceptable.

DISTRIBUTION

The deciding factor on what is the best projector is most often the quality of image that can be created, but this will not be the only consideration for the design team. Two very important factors that can play into choosing any display are control and connectivity. Specific control of displays will be of more interest when designs become more complex, though even simple designs may need to have this as a consideration, depending on where it is installed. But every design will need to consider how to get the content to the display, and this comes down to connectivity and distribution. While some projectors will have user-swappable option cards, most models (and video walls) will have manufacturer-selected connections.

A budget-conscious designer may need to purchase older, used technology. Many of these older displays had a mixture of digital and analog connections. Discussion of the older analog technologies is beyond the scope of this text; modern designers will likely not encounter the challenges of analog-only connectivity. There are enough considerations to take into account with digital distribution, and there are means of keeping the entire signal path with digital delivery.

As there is a variety of codecs with benefits and drawbacks, so there are several standards for transporting content from the source to the display. The standards for distribution are privately held, requiring manufacturers to pay for their use. Most often, this will not impact the designer (as long as the manufacturer does things properly). We will discuss the three most common transmission protocols and the associated cabling, assuming a single signal to a single display. This will be approached again in Chapter 8 as additional equipment may need to be included in the overall design, especially when distributing signal to multiple displays.

HDMI

The most common video connection, due to its prevalence in the consumer video market, is High-Definition Multimedia Interface (HDMI). This proprietary interface for transmission of uncompressed video data and (either compressed or uncompressed) digital audio data is maintained by the HDMI Forum. In addition, HDMI can transport bi-directional auxiliary data, including information about the display. Most likely every display will have this technology (possibly even more than one input for it). Due to its popularity, most every source will also have this means of distribution. So what is it and how does it work?

The first thing to understand is that this system of transmission has gone through a number of revisions as technology has advanced. While the cable connection will generally look the same (there are a couple of variations), the processing in the display and the ability to process the signal at the source may be entirely different. This includes updated cable construction methods for higher bandwidth needs.

HDMI was initially adopted by several manufacturers with the implementation of High-bandwidth Digital Content Protection (HDCP) to prevent copying of digital audio and video content. HDMI was developed initially to be compatible with Digital Video Interface (DVI), a display interface which superseded the popular Video Graphics Array (VGA) connector, mainly in use by computer displays. HDMI cabling was designed to be compatible with high-definition television (HDTV) and utilize a smaller connector, making it more appealing to the consumer market. It also provides the capacity to control certain functionality of displays.

With the initial design (version 1.0) of HDMI, a resolution of up to the VESA specification of WUXGA (1,920 x 1,200) at 60 Hz was possible. This closely matched the limits of DVI. As of version 2.1b, resolutions and refresh rates have dramatically increased, with the ability of displaying 8K60 (7,680 x 4,820 at 60 Hz) or 4K120 (3,840 x 2,160 at 120 Hz) at a bandwidth of up to 48 Gbps. The ability to do this requires every part of the system to be in compliance. This does generally allow for backwards compatibility with other HDMI devices if one or more components are not able to meet the current standard. However, this does mean that the ultimate ability of what can be displayed is established by whatever is the lowest version of the system.

How well and reliably a signal is transmitted depends on many factors – specifically, how a cable is constructed. A limit of only 5 meters (around 15 feet) is achievable with a passive cable (a cable connected to just the source and display without any method of strengthening the signal). Cables of longer length can have video signals degrade through signal attenuation. The greater bandwidth required (due to higher resolution, frame rates, color data, etc.) will reduce the distance a signal can travel without degradation over the same construction cable with lower bandwidth requirements. In a home entertainment system, these limits are unlikely to be reached. However, in a live entertainment venue, even under ideal circumstances, these limitations can be a real concern.

As HDMI may be the best option, there will likely be a need to find alternatives to extend signal distribution. But other challenges can also occur with this form of distribution. As this form of transmission was developed with HDCP, to pass the signal from source to display, all components must be in compliance (part of the signal transmission is a "handshake" between devices informing each other of compliance). If one part of the system is unable to properly transmit its compliance, the video signal will not pass, presenting a completely black image. Diagnosing this issue can be difficult.

As with analog signals, HDMI cabling can also be susceptible to electromagnetic interference (EMI). Later revisions, and cable construction which follows these revisions, limit the amount of interference. This reduces the likelihood of degraded signal quality and artifacts which may become present. The quality construction of these cables will have a significant impact on the cost, making alternative means of transmission (especially due to cable length) a priority.

DISPLAYPORT

The successor standardized by VESA to DVI is DisplayPort. While being mainly developed for monitors and embedded displays, this connection not only carries the video signal, but also audio and other forms of data. It can be compatible with some versions of DVI and HDMI through the use of passive or active adapters. Its transmission protocol relies on packetized data, similar to signals being sent via Ethernet and other computer signals.

Equal to HDMI, DisplayPort is capable of high resolution and refresh rates. Outside of live entertainment, this makes it ideal for such demanding visual tasks as video editing and gaming. In order to do this, it offers high bandwidth capabilities which enable smooth playback of high-resolution videos and 3D content without compromising image quality or performance. For version 2.1 of the DisplayPort standard, transmission rates of 80 Gbps are possible with 8K resolutions at 85 Hz. It has offered variable refresh rates (VRR), which is the key feature to getting the smooth, artifact-free picture, since version 1.2 and only offered with HDMI 2.1. This concept prevents tearing, an error when the display's refresh gets out of sync with the rate of refresh of the source of the content. This is generally seen in fast motion sequences. Most often the extremely high resolutions and data transportation will not be part of the design, so in this case the two technologies should be considered equals.

For live entertainment, DisplayPort offers additional benefits over HDMI. It can drive multiple displays from a single output using Multi-Stream Transport (MST). This technology allows the designer to have an expansive workspace and then that same capability can drive video walls without the need for additional graphics cards. One bonus of MST is that the multiple displays being driven through the single output can have various resolutions. This requires multiple outputs for other technologies.

The cable construction tends to maintain signal integrity over longer distances when compared to HDMI, though often still being too short for most theatrical needs. While HDMI generally relies on a compression fitting (some specialized cables and equipment allow for a screw lock), standard DisplayPort cables feature a locking mechanism that secures the connection between devices, minimizing the risk of accidental disconnections. As with HDMI, the cable standards have evolved over the years, and in order to maximize the capabilities of the latest standard, the source, display, and distribution all need to be made to that standard. Otherwise, backwards compatibility is always an option (though some troubleshooting may be involved when the distribution is the weak link).

While DisplayPort has many advantages, it has its drawbacks as well. The technology is supported mainly by computers and associated display monitors. It is not as widely adopted by all displays. This means that the system designer may need to take into consideration some form of adapter or converter so that the signal can be transported to the display. This can sometimes be a simple passive adapter cable or it can be an expensive converter which could entice the designer to shy away from this technology unless absolutely necessary. The cables designed for the latest standard will also cost significantly more than some other technologies (as do the similarly designed HDMI cables).

As stated, HDMI is more widely adopted, especially in consumer-level electronics. Due to the limited availability in displays and distribution equipment, HDMI may be a better choice for similar capabilities. This becomes more of an issue as systems grow and interoperability in mixed-interface environments will become more challenging.

SDI

Serial digital interface (SDI) is a family of digital interfaces (including HD, 3G-A, 3G-B, 6G, 12G, and 24G) used for broadcast video supporting a variety of resolutions, bitrates, and color depth. These SMPTE video standards are being carried over 75 Ohm co-axial cables. The most common connection for these cables in professional devices is the Bayonet Neill–Concelman (BNC) which maintains the same impedance of the cable. As this transport method was developed for the broadcast industry, it offers a robust video transmission capability suitable for many professional industries, including live events. By maintaining uncompressed digital signals, it is able to ensure real-time signal transport for latency-free (no delay) reliability without compression artifacts or loss of detail.

Compared to HDMI and DisplayPort, which are both developed with the source and display in relatively close proximity to one another, SDI signals can travel relatively long distances without significant signal degradation, making it ideal for large-scale installations. With an appropriate cable, signals of up to 3G-SDI (1080p at bitrate of 2.97 Gbps) for up to 100 meters (over 300 feet). In order to get these distances on higher resolutions, the standards allow for up to four cables in tandem to drive 12G-SDI.

While audio may not be at the forefront of the projection designer's thought process, SDI offers some additional benefits. SDI can carry multiple channels of embedded audio along with the video signal, with the potential to simplify audio video synchronization. Up to 16 channels of audio in HD interfaces are available, and up to 32 (16 pairs) of audio in 3G-SDI and later. The audio data may be useful for advanced designs. This includes the ability to transport timecode, which will be extremely useful in synchronization with other departments.

As with each form of video transmission, SDI has its drawbacks. SDI is compatible with HDMI signal transport and is often used to extend the distance of an HDMI signal through the use of a signal converter. However, SDI does not support HDCP. Due to this, some video signals cannot be converted to and from SDI as it will be a break in the content protection chain from source to display. And as SDI is only available for professional grade equipment, the cost may be prohibitive over similar consumer devices.

SDI has specific bandwidth limitations based on supporting just broadcast standards. HDMI and DisplayPort support not only broadcast standards, but also common computer resolutions. This means that some display devices may not live up to their full potential even if they support SDI transport (such as a projector with a resolution of 1,920 x 1,200 would only display 1,920 x 1,080).

Other methods of digital transport are available to the media designer, though SDI, HDMI, and DisplayPort will be the most common. As the system design comes together, some alternatives may need to be explored, especially depending on the distance between the source and display. We will explore this more in

Chapter 8, Building a system. At that point we will also discuss some additional components which may be required in order to distribute the content.

SOURCE

In the early days of projection design, the source was part of the display – for instance, the film in a motion picture. As systems developed, so did remote sources. For many years, the use of video cassette tapes and optical disc media reigned. These playback devices had very similar consumer and professional devices, making operation fairly simple. However, change was difficult unless each clip used was on its own cassette or disc.

It is possible that a modern projection designer would use older technology, especially if there is no means of transferring it to a modern medium. This is fairly unlikely. For almost all modern designs, the designer will be using some form of media server or a digital media player. These two types of content players offer the flexibility required for constant change and reliability in delivery of content. Yet, the two technologies offer the designer quite different means of doing so in their capabilities and challenges.

DIGITAL MEDIA PLAYER

In the most basic form, this playback device most resembles previous generations of optical disc players. They offer simple playback choices and can reliably skip forward and backward to ensure that the proper video file is displayed. These devices are readily available as consumer devices and can be relatively inexpensive. They are most often limited to a singular output type and may require direct access for control.

A digital media player may contain internal storage, but will often use a solid-state memory device, such as a secure digital (SD) card or microSD card (smaller version of SD card). These flash memory cards were developed for use in portable digital devices. As this form of storage uses non-volatile semiconductor memory to store data files, it can easily be transferred from a computer to the player for content updates. They also allow for easy playback in any order.

As with most computer systems, there are variances in the storage media used for these players. The main variations that should be noted are the storage capacity, speed class, and bus speed. The hardware of the media player should specify the types of storage media along with the file formats to be used. Some of these variations are most important in their write speed, but can have an impact on playback as well.

Digital media players are commonly used in positions where constant video is required, such as theme park rides and digital signage applications. However, due to their small size, they can readily adapt to other video uses. Designers who are on a tight budget and are considering a consumer model player should be wary of limitations in on-screen display of player information. Professional models should have the ability to prevent system information from displaying on screen.

These devices are often limited in formats (audio and video) and resolutions. The output format is also generally limited to HDMI, with the strengths and weaknesses of this type of connection. They are ultimately designed to easily display high-definition images and videos. In the toolbox of the media designer, these players can be quite useful, under the right circumstances.

MEDIA SERVER

The workhorse of the projection designer is the media server. This is a catch-all term for a computer system or software that manipulates digital media for use in entertainment. It often functions similarly to a non-linear video editor, but in real time. There is an assortment of developers of both the hardware and software used in media servers at all levels of price points. The options available in hardware, along with the variations in software methodology, are numerous. How the designer chooses to manipulate content could play a role in the choice of what specific server will be used. To better understand the choices, let us explore first the hardware and then the software and how they are used.

HARDWARE

As every media server program will be performing its tasks on a computer, it is good to understand the components of the hardware and their respective capabilities. Even if no one on the design team will be building a media server from scratch, understanding performance requirements will assist in specifying a prebuilt unit. While most everyone who will be using digital media probably uses a computer of some sort in everyday life, the understanding of the individual components may not have been an issue. This is not a debate between operating systems (Windows, Mac, Chrome, or Linux), though the media server software is likely built for one or two in particular. Instead, we will be discussing the commonalities of the components to include the video capture card, video graphics card, memory (primary storage), file storage (secondary storage), and the network infrastructure.

First, let us explore file storage. The designer needs to be able to keep the content and be able to access it on the computer. Variations of storage will include optical storage, magnetic, flash memory, and network (external). Each form, as with every component of the projection design, will have benefits and drawbacks. Even if the design team is purchasing a prebuilt computer to run a media program, they may have some flexibility in choosing certain components. The decision will be between budget and performance.

The hard disk drive (HDD) utilizes the most ubiquitous form of data storage, magnetic manipulation. This uses a magnetic platter which spins rapidly with a read/write head that gathers/stores information. This is the most common form of data storage and is by far the most economical, especially when compared to the storage capacity.

Flash memory comes in the form of both internal and removable storage solutions. Useful for transporting files between machines (as content is not directly created on the media server), the Universal Serial Bus (USB) drive or memory cards are the most common method of quickly transporting files. While their storage capacities are getting tremendous, some media files may exceed their capacity.

The solid-state drive (SSD) is a replacement of the HDD, but can also serve as a large-capacity portable drive. The SSD is essentially one solid circuit board full of memory chips which makes it extremely fast. It can be more rugged as it has no moving parts; this keeps it smaller and allows some computers to be reduced in overall size. This is especially true with the M.2 SSD, which is not much larger than a stick of gum. If any solid-state media is used as external storage, the speed of accessing the data can be diminished greatly, depending on the method of connection.

On occasion, the designer may need to access files which are not located on the specific device. Consumers may be regular users of cloud storage, whereby files are stored through an internet connection to an outside provider. This storage option is good during initial content creation as the designer may be able to be in a different city than where the performance will be. This makes sharing of data convenient, but is not a solution for the actual performance. Ideally, files should be local to the machine using them. That being said, network-attached storage (NAS) is an external storage solution where devices within a local network can have access to an external storage device (the NAS is an enclosure for either an HDD or SSD) for file sharing as well as file backup or archival purposes. These enclosures can be accessed through either wired or wireless methods.

While secondary storage devices are intended to store files indefinitely, the primary storage is the medium of memory that holds files for short periods of time while the computer is in operation. Random-access memory (RAM) and cache are two examples of primary storage. This form of memory is directly accessible by the central processing unit (CPU) to complete tasks. The amount of primary storage is crucial to the operation of a computer as it holds the directives that the CPU utilizes to perform its tasks. Its proximity to the CPU allows it to provide faster data transfer than secondary storage. Complications where the CPU has insufficient primary memory can include applications slowing down or crashing (failing).

In order to process the visuals in the media server, some means of generating those graphics is required. Two means of doing so are integrated and discrete (dedicated) graphics cards. Integrated graphics are built into the CPU, but are generally lacking in performance. It lacks the necessary architecture to render graphics quickly and efficiently. However, integrated graphics are much less expensive and is dramatically smaller (making it popular in laptops and small form factor desktop units).

On the other hand, a dedicated graphics card is a specialized, high-performance memory device designed specifically for the needs of graphics applications. This is critical for tasks which use significant amounts of graphic processing, such as the demands of a media server. It offloads the graphics processing from the CPU (which is already being taxed with the rest of the tasks involved in the media server) to the graphics processing unit (GPU), which is the specialized processor for dealing with graphics in real time. It has its own primary storage known as video random-access memory (VRAM). In addition, a discrete graphics card can often support multiple video outputs. Graphics cards can be one of the most expensive components in the computer. This helps explain why dedicated media servers can be so expensive and why home-built systems may not be able to compete.

Not all content will be stored on the server. In fact, for some applications, live content from an external source, such as a video camera, will need to integrate into the design. Due to this, many media servers also employ a video capture card. These devices often include both an input for the external video signal and a pass-through so that the signal can be viewed by a monitor at the same time. They process the incoming video stream and will convert, if necessary, into a usable video codec. In a custom-built computer, this can either be an internal card or an external device (transmitting the data through USB).

SOFTWARE

Media server software approaches the handling of content in numerous ways. Some of the considerations will be how the media is organized, whether it is established in linear (sequential) or nonlinear (out of sequence) playback, the manipulation of media, and the level of complexity in operation. Each program will utilize one or more of the following methods of content manipulation. How each software developer approaches these aspects can be quite different. Content can be stored in the same file as the software or as a link (requiring a fixed file path) to some place outside the production suite.

The most popular style of media server for broad use is a cue-stack-based program. These can be simple to use, though provide little in the way of manipulating content. They offer robustness and ease of use. The framework is very similar to how other theatrical systems are organized in the ability to access files and play them with cue timing. This form of media playback relies heavily on pre-production work in the content creation. They offer various levels of video manipulation, mostly in where the video will play (especially with systems using multiple displays) and allowances for mapping (shaping) content onto an irregular surface. This type of program allows for layering video, transitions, and cued video effects. The workflow is fairly straightforward, allowing for novice video designers to get started rather easily. Cue-based systems are often an intuitive choice for integrating show control.

The types of media server software which may make the most sense to the projection designer are those based on timeline control. This style most closely resembles a non-linear editor which is used in content creation. The graphical interface will have a playhead which moves across layers of video, allowing for compositing and sequential playback. The timeline playback is linear, though control cues allow for customized playback sequences, jumping to various points along the timeline. This can be instances of automated cues within the program or triggers initiated by outside sources. Auxiliary timelines can often be created as a feature for asynchronous projects, especially helpful in interactive designs. In addition to geometry correction, timeline-based software often has more advanced features, including advanced live compositing effects and the ability for real-time tracking of physical objects. As more manipulation of the content is happening in the server, rather than previously rendered content, changes per the director can happen much more quickly and effectively than with a cue-based system. The level of complexity in programming increases dramatically over cue-stack systems and thus requires a more experienced programmer to fully take advantage of its possibilities.

A third form of software is based around nodes. A node is a graphical representation of a simple framework of code allowing for a canvas for computational thinking. It allows for a diagram-like dataflow through a representative data-processing interface, interconnecting the blocks of code. It is an object-oriented programming environment. The nodes are often rectangular in shape with points of interconnection. How nodes are connected, often even their placement on the programming surface, will establish a hierarchy of operation. There is not really a predetermined method of use, and the software contains libraries of nodes to provide functionality in highly customizable configurations. For a visual designer, this may be the right fit, as there is a graphical representation of how the various content manipulations work together for the final output. The thought process may seem daunting at first. Many designers see the flexibility of node-based programming as freeing and allowing immense artistic freedom. This is especially true for an experienced user who will be able to create custom nodes for specialized functionality (such as integration with non-theatrical hardware).

While less suited for a theatrical environment, there is an additional style of media server targeted at other forms of live entertainment, particularly live music and arenas, to display emotive and textural content. These are often used by video jockeys (VJs) who work in tandem with disc jockeys (DJs) in their live performances at festivals and clubs. This type of program allows for rapid selection of content and effects in an individualized manner (generally a grid of thumbnail images that can be easily selected). Some versions include the ability to add options to map images onto irregular surfaces or to blend multiple projectors to create larger images. Most will have options to import audio to add levels of control. The method of the workflow is intended to be done live and typically does not have the ability to have predetermined looks in the same way other media servers do. They can be implemented with additional hardware and software to great effect.

OTHER CONSIDERATIONS

The media designer may have many reasons not to use a media server or other media player. Often, this comes down to resources. While some circumstances could require resorting to a different playback device, such as an optical disc player, this is becoming increasingly rare. At the same time, the cost of purchasing or renting a media player system will often be greater than many productions can afford. This is why there are some software-only media server options, allowing for computer hardware that may already be owned. Some of these programs even have the option for rental, just for the duration of the performance.

The creative designer may look to other software options, some of which are offered for free. These will often have much lower functionality, but are perfectly reasonable. The most common programs outside of a dedicated media platform are presentation software, as they offer transitions between images and the ability to cue video playback. At a pinch, an extended desktop background image can suffice for a scenic element. In other words, the designer may need to take some creative liberties and not get discouraged that there is no possibility of creating a design if the normal tools are not available.

BOX 6.4 SUMMARY

- There are three classes of components common to all media design: the source, the means of distribution, and the display.
- Video walls comprise either multiple video monitors set in an array or specialized LED panels.
- Video walls are a direct-view display with the intention of the audience directly looking at the image source.
- Video walls are a good option for when there is less control of ambient light.
- Projectors have a wide variety of technologies to create illumination and the image.

- Understanding light output and the associated contrast ratio is crucial for creating a quality image.
- There are three primary distribution protocols used in live entertainment: HDMI, DisplayPort, and SDI.
- Primary concerns are over bandwidth and cable length.
- The source of content playback can be on a digital media player, but most often comes from a computer running some form of media server software.
- Media servers are either hardware with software or stand-alone programs to be used on the end user's computer.
- The choice of media server takes an understanding of computer components as well as the programming environment of the software.
- There are different styles of media server programs: cue-based, timeline, node, and video mixer style.

SHOW CONTROL

Show control is the centralized management and coordination of various technical components of live entertainment. Specialized components may be necessary, but specifically software will be required in order to automate and synchronize elements of lighting, audio, video, special effects, and scenic movement in order to create immersive and engaging experiences for audiences. Show control systems have been designed to streamline the production process and ensure precise execution of cues.

It is an important part of many productions. Show control provides precise timing along with the synchronization of different departments. This level of precision exceeds human interaction following the call of a stage manager. This accuracy ensures that cues are executed at the right moment, which in turn enhances the overall flow and impact of the performance. The consistency eliminates the variation of operators and human error ("I didn't hear the cue"). This maintains the integrity and quality of the production from day one to the last call.

Modern productions often involve a multitude of highly technical elements and effects. Show control systems aid in the management of the complexity of design by automating the execution of cues and coordinating the interaction between departments and even components within a singular system. This effectively reduces the workload on operators, allowing them to focus on error checking if available. The software can provide feedback and monitoring tools for the operator to check on the status of the cues, devices, and the system in real time. Status reports, error logs, and diagnostic tools can also be employed for troubleshooting and debugging on the occasion that the system does not perform as expected.

For the design team, the use of show control can enhance the production by giving it a more cinematic feel or even allowing the audiences to more easily suspend their disbelief. The coordination of show elements will amplify the impact of the performance on the audience. Through the seamless execution of cues and effects, the audience is less likely to be taken out of "the moment," ensuring a more captivating experience.

Flexibility within the system should allow for adaptation of changes in the production. This flexibility will also allow for the timing adjustments or last-minute changes requested by the director. The system should be designed to accommodate unforeseen circumstances, to guarantee smooth performances under the changing conditions of a live environment (the performing artist may not always hit their mark as expected). Due to unexpected circumstances, the systems must incorporate safety features and fail-safe mechanisms to mitigate risks associated with live performances. While automated shutdown procedures and emergency overrides may be in the area of specialty in the scenic department, there may still be valid reasons for the media designer to consider these possibilities as well. Planning for this type of variation in the performance can aid in the safety of performers and crew, as well as audience members. In addition, user permissions and access control, especially in multi-operator environments, ensure that changes are intentional and expected. These restrictions provide security and safety through controlled access to the system.

By streamlining the production workflows, show control systems improve the overall operational efficiency and contribute to cost savings over time. On large productions, keeping departments separated in order to handle challenges that may arise during a performance is wise. However, smaller productions are often limited in space and ability to adequately staff all operator positions.

By utilizing show control elements, even smaller production companies can provide technically complex shows. These systems are scalable and adaptable, even allowing for touring productions to perform in various-sized venues.

The components of show control will depend on what is necessary. It will require some form of centralized controlling console. There is a variety of communication protocols, which must be observed by both the sending and receiving equipment. For an immersive experience, specialty sensors may be utilized in order to facilitate triggers or other media manipulation. Finally, a well-developed system will include some form of redundancy in order to ensure that "the show must go on."

The projection designer should already be in possession of automation software within the media server. If using other forms of content playback, show control may not be off the table, although some creativity in coordinating this equipment may be more challenging. The cues built into the media server, whatever the style of programming environment, should have means of either receiving an external trigger or of sending commands to other systems. How these events are handled through various protocols is dependent on the individual program.

The facilitation of the cues is executed through various parameters, such as the cue type, the trigger conditions, timing offsets, and various other cue effects. Automation may require advanced custom scripting. Node-based media servers may be best at custom scripting, but even some timeline-based systems allow for custom program elements for the purpose of show control. Common scripting languages are supported to accommodate knowledge levels.

SHOW CONTROL PROTOCOLS

As show control is most often integrating with equipment from different areas of the production, such as audio, lighting, and scenic elements, numerous protocols have been adopted. Many of these were developed as protocols to automate elements within a particular discipline and have been utilized by media servers as tried and true measures of integration. These protocols include

DMX, MIDI, Art-Net, sACN, OSC, or simple TCP/IP. In addition, many devices will support timecode synchronization to ensure the precise timing across the various elements of the production. There are a couple of variations of timecode, such as SMPTE timecode and MIDI timecode.

DMX

Digital Multiplex (DMX) is a standardized protocol primarily used for controlling lighting fixtures and special effects devices (such as hazers and fog machines). It was designed as a de facto protocol to communicate between lighting consoles (also known as desks or controllers) and individual devices (dimmers, intelligent fixtures, etc.) to allow for precise control regardless of manufacturer. In 2004, DMX-512A was adopted by ANSI as the "Asynchronous Serial Digital Data Transmission Standard for Controlling Lighting Equipment and Accessories." It is a unidirectional signal, meaning that the commands go from the controller to the controlled devices, but no signals return from those devices.

The DMX protocol is designed to give instructions per address (numeric designation from 1 to 512 in a given universe, the grouping of 512 channels). Some devices will use multiple channels to control various parameters, but the address will be the starting (lowest) channel. Lighting consoles are programmed with cue timing and values for consistent replication of looks. Media devices began adopting the DMX protocol in order to control functionality from a lighting console. For displays, such as projectors, the control is much more limited than a lighting fixture.

DMX is typically a stable signal over long distances using a cable with a 5-pin (or sometimes 3-pin) XLR connector. It employs a serial communication system which is effective over long distances in electrically noisy environments. However, even with the communication system and the use of a differential signal (signal is interpreted by reading the difference between one data wire and another wire with the inverse data), the standard recommends a maximum cable length of 1,000 feet (304.8 meters). It is also relatively easy to distribute DMX signals to multiple devices. This can mean that there is a reliable signal transmission to all connected fixtures.

As an industry standard, DMX is widely supported by many different manufacturers; it is possibly the most used non-video protocol available in video devices. This gives the design team the greatest opportunity for interoperability and compatibility between different components of the system. For the projection designer, these controls can be utilized for both projectors and media servers. The DMX values can control specific functions of a projector (commonly, the internal shutter, which blocks light output instead of only video black) which can aid in allowing lighting effects to happen in conjunction with projected effects (like a blackout).

As DMX is a unidirectional protocol, it is unable to receive feedback from projectors or the media server. This prevents the ability to receive status updates and error reporting, which could prevent necessary feedback. This is especially difficult when depending on DMX to be transporting signals over great distances. Signal boosters and repeaters can extend the range, but may also introduce unpredictable latency. Without bi-directional communication, troubleshooting and diagnostics will become a challenge. This is of particular concern as high levels of EMI may cause signal interference and lead to erratic behavior.

While DMX remains a prevalent and versatile control protocol, its limitations should be considered when implementing show control with it. As a media server can be controlled by the lighting console in a number of ways (modern consoles have multiple control protocols beyond just DMX), it would be prudent to consider all options. Many lighting designers choose alternatives to DMX for similar reasons to those of the projection designer in wanting to consider other protocols.

MIDI SHOW CONTROL

Musical Instrument Digital Interface (MIDI) is a communication protocol that was developed for musicians in order to control a variety of musical instruments and related audio devices through digital interfaces and specialized electrical connections. Each connection can contain up to 16 channels of information. It began as a set of standardized codes that represent various musical parameters such as pitch, tempo, volume, and so on. As it is a protocol, and not an actual musical waveform, it is a form of computer language that triggers commands. While it was developed in the early 1980s, it is far from obsolete.

Through its evolution, MIDI expanded to control lighting, trigger special effects, and provide automation. MIDI data contains essentially two types of commands – notes and controllers. The note data sends a command when the note is "played," when it is released, and the velocity at which it was initiated (it may take a specific type of control surface for this).

MIDI utilizes a 5-pin DIN (Deutsches Institut für Normung, the German Institute for Standards) connector. Similar to DMX, this is a unidirectional protocol. When bi-directional communication is required, a second cable is necessary. The standard maximum for cable length is about 50 feet (15 meters). And as most devices do not copy their messages from the input to the output port, a third "thru" port emits a copy of every signal received on the input port. Some devices also lack the ability to generate MIDI, and will not have an "out" port as a result. Planning on how devices are connected is required. In addition to the protocol advancing, so too have the controlling devices.

About ten years after MIDI was developed and widely adopted, a new standard to expand beyond just musical instrument control was ratified by the MIDI Manufacturers Association. MIDI Show Control (MSC) is a protocol standard allowing for a range of entertainment control devices to communicate in order to perform the tasks necessary in entertainment applications. Like the original MIDI protocol, it is a control protocol and does not contain any show media. The protocol is similar to the original MIDI protocol and is thus compatible with conventional MIDI hardware, including control devices. Many MSC devices use Ethernet communications for expanded bandwidth and the flexibility of that networking structure. In addition, as computer control is common, MIDI is also transmitted via USB connections.

Use of MSC allows for precise synchronization of various effects (lighting, video, audio, etc.) through MIDI timecode or MIDI clock signals, regardless of manufacturer. In order to minimize latency in transmission (a result of daisy-chained devices), additional management devices are available, providing multiple thru ports, allowing multiple devices to be connected to a single control surface. This allows for the precise timing needed to trigger the pre-programmed cues without the need for manual intervention. Although if changes are required during a performance, operator control over show elements is still possible through manual adjustments.

As MIDI is limited in the number of channels of control, complex shows with a multitude of elements requiring show control could easily exceed its transmission abilities. For simple control requiring sequential triggering, this may be ideal. However, with the limited commands available, advanced video effects may be beyond its limited functionality. This should be at the forefront of the designer's mind as it will have limited scalability. If control systems have the potential to increase in complexity, potentially exceeding the limitations of MSC, then other protocols might need to be explored early on, so as not to bottleneck the production schedule while new show control options are implemented.

Overall, MSC offers a flexible and cost-effective solution for the synchronization of show elements. It has its limitations in functionality and scalability in comparison to other show control protocols. The MSC syntax system can seem quite complex, especially when unfamiliar with the concepts of configuring devices to properly use MIDI commands, which requires a strong understanding of its methods for quality enactment. However, with careful planning and implementation, the effectiveness and reliability of a MIDI show control system can be the perfect fit not only for the media designer in coordinating the video elements, but with the production as a whole.

NETWORK CONTROL

The Internet Protocol (IP) is a suite of communication protocols developed with the intent of networking devices on the Internet (a global system of networks) as well as private networks. For the majority of installations, the designer will only need to worry about how this method of communication works within a limited space like other show control elements; this does offer flexibility to offer remote performance options as well. IP obtains and defines an address, which is a designation where data must be sent.

Ethernet, the backbone architecture of local networks, connects devices within close proximity efficiently and reliably. This protocol for a local area network (LAN) is the framework to facilitate communication between computers and other devices within a limited area. It allows for ensuring fast and secure communication between the devices. Most data is transmitted over twisted pair cables (standards will vary and are categorized, e.g., Cat5, Cat5e, Cat 6, etc.), though wireless transmission is also a possibility.

Wired communication is generally preferred for reliability, though the choice to go wireless depends on the specific needs and constraints of the LAN, including range and bandwidth requirements, as well as the physical layout of the space (which could limit wireless transmission). Transport speeds in a LAN can be quite high, generally supporting all video content and show control needs. The bandwidth requirements for video can be quite taxing on some systems, especially when older equipment is still in use, causing choke points. Wired communication will generally have substantially faster speeds as well.

Whereas DMX and MSC utilize serial communication, IP transmits information via packets (compartmentalized messages). This reduces the size of messages and reduces bandwidth requirements if there is a need to resend data. The packets are reassembled at their destination. Each packet can take a different route between the source and destination in case one route becomes congested (reduced transmission rates) or completely unavailable. There are three distinct methods by which network data can be delivered: broadcast, unicast, and multicast. Broadcast delivers all data to all devices without exclusion. Unicast uses the unique address of each device to deliver specific data to the intended device. Finally, multicast is a method in which devices subscribe to the data they require. This could be considered as relationships of one-to-all, one-to-one, or one-to-many, respectively.

The Transmission Control Protocol is the data transport protocol which includes strategies for packet ordering, retransmission, and data integrity. TCP solves many issues that can arise through transmitting data in packets, such as lost packets, packets arriving out of order, duplicated packets, and packets which become corrupted. As this protocol is used on top of IP, it has become known as TCP/IP.

An alternative to TCP is User Datagram Protocol (UDP) which does not provide error correction or packet sequencing. It also delivers data in a method which does not designate the receiver. For these reasons, it offers lower latency between connected applications by decreasing transmission times. In some ways, this gives it an edge over TCP for time-sensitive communication (such as streaming media). However, as there is not a component of error checking, this protocol is a poor choice when interfacing with automated scenery.

Why would a production choose network protocols for show control? As can be seen in the "Internet of Things" in consumer devices, this form of communication is pretty much everywhere. This means that a wide variety of entertainment devices are likely already enabled to be networked, some even with wireless communication potential. This allows for ease of integration and control of elements within the show control system. This also permits the possibility of connection with external systems and services (including cloud-based media storage). Scalability and flexibility are thus possible, allowing for expansion or fine-tuning or to easily accommodate changes in production requirements.

Unlike previous serial control communication, network protocols can support monitoring and diagnostics of the systems, through device status, error logs, and potentially other performance metrics. If the system is connected to an external system, remote troubleshooting may also be possible. In addition, the variety of equipment to be used in a networked show control system is readily available. In particular, at least one manufacturer has created a gigabit switch specifically with the entertainment industry in mind, for the distribution of lighting, sound, and video signals.

Things which need to be considered when using network protocols for show control will include: network reliability, security, and whether the complexity can be met. Any network issues, such as latency, packet loss, and network congestion, can disrupt essential communication between devices and compromise the stability of the system. Reliability may be of particular concern when unauthorized access is possible. This comes into play especially when wireless communication is part of the system (as it is easier to secure wired communication). Malicious individuals can cause at best disruptions and at worst safety hazards. Thus, the complexity required in a system (including proper network security measures) may be beyond the ability of show personnel. Even bringing in talented information technology (IT) staff may not solve every issue. Configuration of networking elements, as required by certain show control elements, can be different from what is generally required in a traditional office environment. As a final note, it is important to create a dedicated network structure and not attempt to share the network being used for non-show purposes.

ART-NET

A specialized network protocol, initially designed for transmitting DMX and remote device management (RDM), Art-Net uses UDP to communicate between devices. It has become a widely used protocol in the entertainment industry. It offers the same advantages of DMX communication with the advantages of Ethernet-based communication (including bi-directional communication). For departments outside of lighting and electrics, it also allows for the use of commonly available cables, instead of having to invest in dedicated DMX cables. A wide variety of hardware is available to work with equipment not directly manufactured to work with this protocol.

For show control, Art-Net is able to include management functions such as detecting and controlling devices (generally an output device such as an intelligent light or projector) as well as transmitting timecodes. It is scalable as it supports multiple DMX universes (up to 32,768 unique numbered universes) over a single connection. This can reduce the complexity of larger systems in data distribution.

Art-Net is developed by Artistic Licence Holdings Ltd., which has made it open for use on a royalty-free basis, and is supported by a wide range of lighting consoles, media servers, entertainment fixtures, and other show control devices from various manufacturers. This ensures interoperability and compatibility between different components of a show control system. Its seamless integration with media servers to synchronize lighting effects with multimedia content is widely adopted, especially by large music events.

As with any network-based protocol, Art-Net relies on a reliable network for communication between devices. As not every portion of the protocol receives return packets, including the ArtTimeCode, packet loss or network congestion can disrupt the transmission of data and compromise the stability of the control system. While it supports an extraordinary amount of universes, the network bandwidth may limit the amount of universes that can actually be transmitted simultaneously, especially when using the broadcast method. Addressing of connected devices is limited to two structures: 2.xxx.xxx.xxx or 10.xxx.xxx.xxx. With proper network planning, configuration, and maintenance, the reliability of an Art-Net control system could be the right choice. A free troubleshooting software application for Windows computers is available for download (from https://art-net.org.uk/resources/dmx-workshop/) in order to diagnose and configure the network.

sACN

Another network-based lighting control method which can be adopted as a show control protocol is streaming Advanced Controller Network (sACN). This is an ANSI standard protocol (E1.31), which is a subset of the full ACN standard (E1.17), developed by the Entertainment Services and Technology Association (ESTA) as a result of the need for additional universes required to control a large number of RGB LED fixtures. While the scalability of Art-Net is massive, sACN can support the transmission of up to 64,000 unique numbered universes and can operate on any network address (including mixed addresses on the same network). Devices can use automatic IP address assignment such as Dynamic Host Configuration Protocol (DHCP), which can simplify configuration.

sACN is another DMX over Ethernet protocol which distributes multiple universes of DMX data. Similar to Art-Net, sACN uses UDP for transmitting DMX data packets, allowing both to coexist on the same network. However, while some methods of Art-Net utilize broadcast methods, sACN primarily uses unicast or multicast transmission, which can reduce network congestion.

As many network infrastructures are designed with the expectation of using TCP/IP traffic, they are also designed to expect the majority of data to be unicast. With entertainment control networks using a significant percentage of UDP, some network products may see an abundant amount of broadcast data as a fault and dismiss it. This is a feature known as broadcast storm protection.

Network switches (a device which forwards data to connected devices through packet switching) have differing levels of support for multicasting (the preferred method of sACN). The switch designates which multicast subscribers are connected to individual ports through the monitoring of Internet Group Management Protocol (IGMP) packets. Errors in multicast can include the switch not receiving IGMP packets and thus blocking multicast data or converting the packets to broadcast.

Multiple controllers are allowed to be used with sACN, assigning a priority to the controllers. This can include standard situations where two control desks can be set up as a main and a backup (should the main fail). However, split control is also a possibility for certain show elements. For instance, the lighting console may have a lower priority on lights which are pixel-mapped (lighting fixtures being used in an array to replay images or video) than the connected media server. Thus, the lights would follow the directions provided by the lighting console as long as no commands were being provided by the media server.

While sACN may sound as if it is the hands-down superior protocol, there are some things to consider. Art-Net is more widely adopted (developed sooner for one) and is supported by a greater number of devices. In addition, Art-Net is able to use RDM natively, while sACN would need to be combined with RDMnet (ANSI E1.33, Message Transport and Device Management over IP Networks) to match in comparison. However, sACN should typically be an easier network to set up, mainly due to how IP address assignment is completed.

OSC

Open Sound Control (OSC) is an encoding method (not quite a protocol) which provides a more flexible alternative to MIDI; again, it was initially designed for communication for musical performance. This digital media content format for streams of real-time control messages is optimized for modern networking technology, as opposed to the proprietary system used by MIDI. It offers lower latency, is extensible, and is more suitable for a wide range of applications beyond music and sound, such as show control. As it is based on standard IP networking protocols, it is suitable for use with a local area network, including wireless communication.

OSC messages utilize a Uniform Resource Identifier (URI) style of symbolic naming scheme. This is a syntax method used by the World Wide Web initiative. This allows for hierarchical organization of the address space in human-readable patterns. The transmission of messages containing arbitrary data types allows for flexibility and precision over various parameters of show elements. OSC streams are sequences referencing a point in time called a time tag. The stream sequences are bundles of data, which contain a number of messages. Each bundle of messages represents the state of the stream at the respective time tag. Bundles may contain sub-streams which can be sampled at various rates.

Similar to Art-Net and sACN, OSC is commonly transported using UDP. However, it can be encapsulated in any digital communication protocol to the benefits and drawbacks on quality depending on the method chosen. While using UDP, OSC bundle timestamps can be used to correct packets which may arrive out of order. This obviously can introduce extra latency, especially if the packet is fragmented over multiple pieces.

Assured transport of data, which may be necessary depending on the elements being controlled, should be sent using a serial stream transport such as TCP/IP. This provides a continuous data stream between end points. As OSC bundles can contain a variety of encapsulated messages, extra data is required to indicate packet boundaries so that they are not divided.

OSC is supported by a wide range of entertainment software applications, providing the designer with relative ease of utilizing it as a method of control. Hardware, including lighting consoles, media servers, and many audio devices can all communicate with each other using OSC messaging, allowing for flexibility in design and operation. The customization (including the ability to use custom control interfaces) allows for commands and data formats to be tailored to a specific show control requirement. This can also extend the functionality of existing show control systems for unique artistic visions and technical requirements.

Of course, with the flexibility of OSC comes complexity of implementation. This is especially challenging for users with limited programming and networking expertise. A solid understanding of OSC messages, addressing, and routing will require a good understanding of OSC syntax and network choices. Comprehensive documentation and tutorials are considerably difficult to come by, increasing the difficulties in troubleshooting and debugging OSC implementations. In addition, while custom solutions are available, this may require additional custom programming and scripting. This is in contrast to MIDI, which has a well-defined set of commands, making it easier to set up show control. Despite these challenges, OSC remains a powerful and flexible show control option.

TIMECODE

In order to precisely coordinate the timing of different elements within a performance or production, a synchronized signal known as timecode is used. This consists of a series of sequential numerical values that represent specific points in time. The film and television industry utilizes a standard known as SMPTE timecode which is represented in the 8-digit format HH:MM:SS:FF (hours, minutes, seconds, and frames) to represent the breakdown. Thus, the actual frame rate of video playback is critically important.

The function of timecode is to provide an exact positional reference. Timecode is generated by numerous professional video recording devices, in order to synchronize clips for editing. For live entertainment, a timecode generator or timecode software applications can be used, usually set to either 24 or 30 frames per second. There can only be one source of all timecode in a system. Once timecode is detected and aligned, the receiver starts moving its own transport to the time that is read from incoming timecode.

Timecode is most often used in musicals for seamless synchronization between show elements. In a theme park setting, a show such as the Enchanted Tiki Room in the Disney parks requires audio, lighting, and the movements of animatronics to be in perfect sync so that the park attendees believe that the birds are actually performing.

Timecode receivers "chase" the incoming signal. In order to determine the way that the device locks onto the incoming signal, the designer may have a few choices within their device. These chase modes start with full sync lock, where the receiver behaves in the same fashion as the timecode signal. If the timecode stops, the receiver stops. This does allow for immediate restart of playback when the timecode resumes, depending on the cues built in. If timecode is subject to drift, freewheel mode may be a better alternative. This allows the receiver to abandon timecode to avoid random stopping and starting if there are dropouts in signal. There is some fault-tolerance which should be employed to make this effective and still provide the benefits. Similarly, if timecode drifts, the chase relock mode can determine whether the receiver shifts to a correct timecode value or whether it is to continue on its current pace. This is most often used in broadcast situations. Which mode is used may depend on the individual receivers and not be universal across the production elements.

LTC

If timecode is used and synchronized to the audio department, it will likely use linear or longitudinal timecode (LTC), which is an SMPTE timecode transmitted as a standard analog audio signal (tone). This is recorded alongside other audio tracks in the playback system. This will have any given frame number corresponding to a very specific point in the track, allowing audio to trigger corresponding show-controlled elements. This audio track will have a separate output on the audio interface. The track will not be understandable by humans if played through speakers.

As an audio signal, LTC can be distributed through standard audio wiring, including distribution amplifiers. It can even be passed over video cables, though some attenuation can occur due to the difference in resistance between audio and video cables, which may cause the signal to be unusable by some equipment. In addition, LTC can have some challenges introduced if an audio processor is introduced with processes such as noise reduction, equalization, or any type of compressor. As LTC is an audio signal tone, it must be playing for other devices to understand the signal in order to lock onto it. The information is inaccessible if the timecode is not in motion.

MTC

As a simple measure for synchronizing musical elements, MIDI is well established. In addition to commands triggered manually by a MIDI controller, the MIDI clock can be used. This is a measure-driven pulse clock that carries Play, Stop, Forward, and Backward commands along with the tempo data. However, for show control purposes, the use of MIDI timecode (MTC) allows for timecode information to be transferred across MIDI in order to allow specific timing of MIDI events to occur. As various components utilizing MIDI cannot read SMPTE timecode directly, the sequencing of MIDI synchronization will require conversion. This conversion is often to MTC, which can communicate directly with MIDI devices.

MTC follows the same format as SMPTE timecode, in the format of HH:MM:SS:FF. It is more advanced than the simple MIDI clock which does not include any positional information, which is important in case some sync gets corrupted or lost. As MTC is transmitted on the same data transmission as other MIDI messages, it is broken up into eight separate quarter-frame messages. This means that it takes two frames in order to provide positional data. Because of this, timing jitter is possible if a MIDI data stream is running close to capacity, introducing latency in the MTC data stream.

As a show control system, MTC is a common method of synchronization. It is distinct from other MIDI messages. MTC is transmitted as a series of MIDI system-exclusive (SysEx) messages that contain timecode data. Most devices which are established to receive MIDI data can also support MTC. This is used alongside MSC and other MIDI-based protocols to create the integrated show control systems that bring the show elements together.

> **BOX 7.1 SUMMARY**
>
> - Show control provides synchronization between production elements.
> - Show control systems include a controller, a standardized protocol for sending and receiving data, and should include elements of redundancy and fail-safes.
> - Common protocols for show control include DMX, MIDI, Art-Net, sACN, OSC, and simple Ethernet protocols.
> - Show elements can be synchronized through timecode.
> - Some widely accepted protocols have limitations which may not be useful in a very complex system, but should suffice for the majority of designs.
> - Each protocol has its intended design which may implement challenges depending on elements to be controlled.

8

BUILDING A SYSTEM

As we come to the conclusion, we again start at the beginning. Like the ouroboros, the serpent which eats its own tail, the eternal cycle of where to begin and end in the creation cycle is ever-present and is always connected. Understanding how the content will be displayed can be as impactful as the content itself. Knowing what will be provided and how it should look is vital to designing the projection system. This is, of course, dependent on the designer coming into a venue where the system is not already installed, limiting the content to be displayed in a predetermined configuration.

Obviously, how the designer approaches the system design depends on the magnitude of the production. If the designer is managing a team, then the designer is more likely to be communicating needs and putting the final stamp of approval on the delivery system, not being concerned with the minutiae of the entirety of the system. On the other hand, for more intimate productions where the designer is wearing many hats, the understanding of the entire system will become much more important.

GATHERING NECESSARY INFORMATION

Designing for any performance requires a bit of homework on the venue where it will be performed. Certain elements of information

DOI: 10.4324/9781003107019-8

gathered will be required by the projection team, while others are required to share among other departments in order to realize the design. While certain information can be gleaned through architectural drawings, it is generally best if the designer is able to visit the space in person, at least once, prior to starting the design process.

What kind of information needs to be gathered? Some assumptions need to be made for the sake of this exercise. The first assumption is that the upcoming production is in a predefined space (even if in a flexible theater, that audience and performance spaces are defined). Second, it is assumed that the initial concept of the production has been set (initial reading of the script, at least the first meeting with the director and other designers to discuss concepts). At least a general idea of the budget should be understood as well as what can be provided by the venue (installed or not).

On the initial site survey, make sure to get contact information for essential venue staff (there will probably be follow-up questions), including the venue manager or technical director, any IT personnel, the house electrician, and possibly the house rigger. Indicate to the production management team if there are specific areas of the venue that will need to be accessed and if any special considerations are required for that access. Essential tools for the design team will be a note-taking device (preferably something to sketch, not just type or audibly record notes), a camera (make sure it can take quality pictures in low light), a measuring device (preferably a laser range finder and a tape measure), a light meter, and a flashlight (small and bright). It is good to have a basic understanding of what the end result of the design will be, to be able to gather pertinent information. This includes whether any of the design will include components of text or other data which must be discernible by the audience (this includes translations).

It is important to discover what the sightlines will be. For the audience, the designer needs to consider a few items for show-critical visuals, including the angle of view, the greatest distance from the projection surface to the farthest audience member, the total extent of viewable surface, and an idea of where the projectors can be placed (this is going to be visited a number of times). If images are purely aesthetic and are not providing show-critical information, these steps can be looked at with a much less critical eye.

> ### BOX 8.1 BEST PRACTICE: CABLE PATHWAY
>
> Take note of existing wiring. The likelihood of the entire required infrastructure being available is slim at best. Thus, take special note of any cable raceways and wall pass-through ports (mouseholes) as well as potential trip hazards for when installing power and signal distribution cabling.

In live entertainment, we often conceal items that could be distracting to the audience. This means that the architecture and added scenic elements will be placed within the view of the audience to minimize the view of theatrical elements that are not a part of the story. These elements could potentially interfere with the projection. Thus, sightlines of the projection equipment must also be considered for any obstructions (current and proposed). Take measurements (if possible) from projector mounting positions to where the surface will be (a temporary surface can be used). If at an angle (most likely higher than the intended image), the measurement should be at the closest point. Again it needs to be emphasized, communication is crucial, especially in the information-gathering phase. Very often only one member of the projection design team will be able to visit the venue prior to production (and sometimes no one at all). It is extremely important to be specific in the requirements necessary for a successful site visit.

> ### BOX 8.2 BEST PRACTICE: STRUCTURAL CHALLENGES
>
> Check the structure for any vibrations where the projector will be mounted. Vibrations can make images appear as if they are out of focus. The proximity of air-conditioning systems is often a source of vibrations. In addition, if the air is directly blowing onto

> a projector or projection surface, this can have an impact on the images produced.

One final note of information gathering is actually for the production team. As some portion of the design will likely be called by the stage manager, the designer should be in constant contact with their team. The designer may need to be kept abreast of blocking notes. If interactive, the designer may need to be present at rehearsals in order to have the performers work with these elements (great opportunity to work out bugs). And finally, the designer will need to know how cues will be called in order to prepare for building the cue sheet.

FRONT OR REAR PROJECTION

Is the projection going to be front or rear projection? Front projection is the most common, where the image reflects off the surface. Rear projection requires a projection surface to primarily pass light through the surface, but must have enough opacity for the light to be viewed. Each technique has its own advantages and disadvantages. Each will have its own unique requirements which the designer will need to consider in the venue.

Let us consider the less common first. There are several things to know when choosing rear projection. First, a certain amount of light is lost as it is reflected back towards the projector. However, the advantages are many and will likely be a tempting choice. As the light passes through the surface, it resembles a seamless video wall. It is a vibrant image when done correctly. If the projector is not able to be directly on axis, slight geometry issues can go unnoticed by the audience due to scenic elements which can block any overspill. Unfortunately, due to material sizes, very large projection surfaces will require a seam, which will be visible in a rear projection setup.

The first thing to consider is that the projection operator will likely be sitting in the front of house, with a clear view of the stage, similar to the lighting operator. This means that the distance to the projectors will exceed many traditional forms of distribution. This can be the case even when not choosing rear projection; so alternative means of signal flow will be discussed shortly. In addition, the viewing angle (degree that the audience is viewing, separated from the perpendicular of the center of the surface) is often less than that of front projection. This isn't necessarily an issue if the audience is mostly seated in front of the surface, but the image will become much dimmer for those viewing off to the side.

The other issue is that the space behind the projection surface may need to be excessive, greatly minimizing the use for crossovers or scenic storage. Backstage space is often at a premium, especially directly upstage in a proscenium space. To create an image that fills an entire backdrop may require multiple projectors, as each one would not need to be back as far as necessary if using a single projector to get the same size image. A single projector with a short throw lens will often need to be directly centered on the projection surface as any deviation will cause portions of the image to be out of focus due to the shallow depth of focus. There are some exceptions as many home and office consumer projectors have lensing to have them very close to the projection surface (limited lumen output). In addition, the designer can use a first surface mirror (where the reflective coating is at the top of the glass as opposed to the reverse side) to have the projector close to the rear projection surface and have a suitably large image. This was a common technique used in the production of rear projection televisions.

A concern with rear projection is that there can be uneven brightness across the image. A projection surface with a greater light transmission will be subject to "hot spotting" (the center part of the viewed image is brighter than the outer radius). A surface with lower transmission will appear to have an even brightness across the image, but much of the light is reflected back, reducing the amount of light reaching the audience, but also increasing the viewing angle. There are some tricks to reducing the hot spot if a wide angle of view is not required and maximum light output is needed. Primarily, the projector needs to be not in a direct line of sight (if there were no surface between the audience and projector). Obviously, with a short throw lens, this comes with all other issues of not having the projector directly on axis with the surface.

Most designs will require front projection. This requires the light to reflect off the surface to be transmitted to the viewer. As this requires reflection, ambient light is of a greater concern than it is with rear projection. Thus, rear projection generally has a better control over the actual contrast ratio than front projection. However, front projection has a greater amount of surfaces available to project upon in general.

The two main concerns of front projection are ensuring that the projected image is bright enough to be clearly seen (poor contrast due to ambient light being overcome by brightness of image) and the shadows that may be cast onto the image, in particular by the performers (see Figure 8.1). As projecting a backdrop is a common use of projection, this comes into play often. The designer has a couple of options to reduce projection onto obstacles. As seen, increasing the height of the projector so that it can project over the heads of the performers is a possibility, though the angle may become quite steep in order to prevent shadows completely. This introduces the geometric distortion known as keystoning. Another option is to choose not to project all the way to the floor and keep the projected image above head height.

The distance from the operator to the projector may be well within distance specifications of many distribution methods when

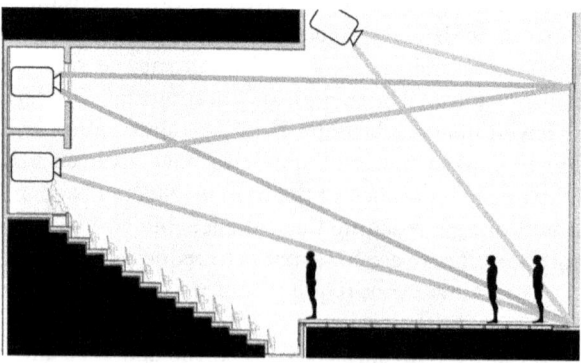

Figure 8.1 Three potential projector positions: the design must face the balance between geometric distortion and shadows of the performers.

using front projection. This is by no means a guarantee. Just as with rear projection, alternative means of distribution may be required, depending on where the projectors need to be placed. The designer should look for alternative distribution over an alternative projector placement. We will discuss some of these alternative means of distribution shortly.

PROJECTION SURFACE

While an LED wall has significant advantages of brightness and contrast due to direct viewing, it cannot match the organic nature of a projection system. Light is predictable and malleable in the means of covering even complex geometric shapes, allowing a talented media team to bring the inanimate to life. There are a few things to understand about the properties of the projection surface that the designer should take heed of in order to choose, or recommend to the scenic designer, the materials and their treatment.

Most materials that are not specifically designed for projection will have unknown properties and will require testing and experimentation to understand how they react (reflect or transmit the image). A projection screen will have some known properties, allowing the designer to understand specifics on how it will react to the projected image (brightness, contrast, viewing angle, etc.). The projection surface is a passive component of the projection system, but plays an integral component for the audience.

To understand how bright an image will be, a property known as gain is listed by the manufacturer. While this is a reflective property for front projection screens, it is sometimes referenced for the transmittance of rear screen material as a comparison to a similar front projection screen. Screen gain is a mathematical representation of how much light will be returned to the viewer. This means that if a material is listed with a gain of 1.0, the same amount of light delivered will be reflected back. If a material has a gain greater than 1.0, it is essentially acting as a focusing agent, making the screen seem brighter than the amount of light being delivered (for instance, a 1,000-lumen projector on a surface with a gain of 1.1 would appear the same as a 1,100-lumen projector on a surface with a gain of 1.0). This does not mean that the designer should always choose a material with a high (greater than 1.0) gain. The higher the gain, the more likely the material will be subject to hot spotting (where a portion of the image is noticeably brighter to the viewer than the rest of the image) due to specular reflectivity.

A property which is often directly inverse to the measurement of gain is the viewing angle (sometimes referred to as the viewing cone). The measurement of gain is taken with the projector perpendicular to the center of the surface. At the angle of view where the brightness drops to 50 percent (half gain) is where the viewing angle is determined (doubled due to that angle being replicated both right and left). It is important to note that the perceived brightness of the viewer will not be half as bright, as our visual system is logarithmic, not linear. To have a very wide viewing angle, the reflected light is being dispersed to a greater extent, reducing the amount of light returning to the viewer. It should be noted that, with projection screen material, there can be two viewing angles, horizontal and vertical (vertical angle not regularly measured). This may have an impact on venues with a balcony, including arenas.

There is another property which can impact the gain of a projection surface. This is that the color of the material will be a shade of gray with deeper shades even approaching black, as opposed to the traditional white surface. The purpose of the colored material is to improve perceived contrast. This will help particularly in situations with a greater amount of ambient light. The more ambient light present, the darker the material will be in order to maintain the perceived contrast. This means that if the same 1,000-lumen projector is on a gray surface, reducing the gain to 0.8, then it will appear to be as bright as an 800-lumen projector on a 1.0 gain screen. Thus, to get the brightness of a 1,000-lumen projector, the designer would need to specify a 1,250-lumen projector.

A final property to consider is that not all surfaces are perfectly smooth. In the case of a projector screen, this texture coming from various coatings or actual physical texture can aid in distributing light in predictable ways for viewing angle and gain. The material may also be perforated, as is the case for acoustically transparent screens or those intended for outdoor use. Projection screens may also be deliberately textured to aid in the rejection of ambient light. This is more common in classrooms or conference rooms where ambient light, especially from overhead lighting or nearby windows, is difficult to control. It is unlikely that a designer would choose this type of material in entertainment as it also limits the position of the projector and audience.

The texture of a projection surface will have additional ramifications depending on the proximity of the audience to the surface and the pixel density being displayed. With lower-resolution projectors, or where the image is rather large, reducing pixel density, the texture of the surface has little impact on the quality of the image. However, if the audience is going to be in close proximity to the projection surface, often eliciting the desire for a greater pixel density for clarity, the texture can have significant impact on the quality of the image. A pixel-dense image may start to appear grainy, out of focus, or appear as if the video has artifacts, all due to pixel distortion.

A projection surface can be made of pretty much anything that scatters light to some degree (more on this in a bit). Projector screens are typically made of vinyl, high-grade plastics, spandex, or polyester fabric; on occasion they may be made of PVC, canvas, or fiberglass. These are treated with a coating to enhance visibility of light (which does mean that special care is required if cleaning is necessary, to prevent the removal of this coating). Front projection screens may even have a black backing in order to prevent light transmitting through from the back which can reduce perceived contrast as much as ambient light from the front.

For live entertainment, a projection screen is often not a viable option for a variety of reasons. The primary reason for not using projection screen material for most live entertainment is that it is a "one-trick pony." While it is really good at what it does – displaying the projected image – it typically needs to be used strictly as that. With the exception of a concert where the surface will have an image on it for the entirety of the performance, other shows, from spectacles to straight plays, will have variations in scenic elements, not requiring constant projection. A projector screen when not in use often looks like just what it is – a piece of material which is lacking something on it.

Since portions of the set will not require constant projection they may need to be painted or otherwise treated to blend in with the rest of the scenic elements. The surface may just be standard scenic materials such as curtains or the cyclorama. Not using a projection screen material will likely have a dramatic impact on the effective look of the projected image. As the surface will not be treated to directly reflect light, the gain is unlikely to approach unity. A general rule of thumb is that most light-colored surfaces will reduce the gain by about half. This is due to the material both absorbing some of the light and increasingly scattering it (giving it a broad viewing angle).

When using any projection material, traditional or not, the designer may need to pay close attention to the color of the surface. As color was mentioned previously in regard to assisting contrast ratio, projection surfaces may also have additional tints. For traditional materials, this is not typical, with most being white or gray, but some specialty screens are designed to enhance certain color spectrums. However, when using non-traditional materials, there will most likely be some color bias. This may be even more evident depending on the projector being used. Each projector will create white differently, depending on the light source and the imaging technology. This is where early testing should be practiced if possible, especially if color accuracy is non-negotiable.

BOX 8.3 SAFETY CONCERNS OF PROJECTION MATERIALS

Non-traditional materials should still meet theatrical standards, specifically fire retardancy. The projection designer will often play with a variety of materials to get different effects with the projected light. Sometimes the budget-conscious designer may need to find a material that closely resembles a traditional projection surface, including other plastics like shower curtains. In the United States, the National Fire Protection Association (NFPA) has two different standards (701 and 705) for testing textiles and films for use in entertainment venues. Though the projection designer may not be well versed in these standards, getting assistance from the scenic designer or technical director may be in order.

There are several choices that a designer might consider with which to experiment as a method of creating a special effect. In order to make the projected image ethereal, some form of atmospheric such as theatrical smoke may be chosen. First, it must be determined whether atmospheric effects are allowed in the venue; then the designer will need time to experiment with controlling the particles to provide adequate material to appropriately disperse the light to be seen as a discernible image. Too much air flow may disperse the atmospheric to the point that only beams of light are noticeable, while too much may have the same effect. Projecting into theatrical smoke is best done as rear projection. However, as most of the light will actually pass through the smoke, the designer will need to understand where the remaining light will go, particularly so it is not distracting to the audience.

Similarly, specialty theatrical fabrics, such as sharkstooth scrim, are popular choices. These primarily need to be subject to front projection for the audience to best see the image. Again, a considerable amount of light will pass through these materials. In this case, the remainder of the light is likely to be seen by the audience. This can be used to the advantage of the design, such as giving additional volume to clouds. However, to get the greatest amount of light on the surface, and not pass through it, the projector may need to be placed off-axis (not perpendicularly) which can create additional image challenges.

BOX 8.4 CASE STUDY: MAKING THE IMAGE WORK

Equally important for what is intended to be seen is hiding what is not intended to be seen. The choice of surrounding and backing materials can impact the perception of the image. There are a number of theatrical fabrics used with the intent of absorbing light, such as duvetyn, with the short velvety nap. The direction of the nap can make a difference in the amount of light it captures. For one project, where the director was looking to project onto a scrim and then reveal a performer behind by lighting them up, they ran into a couple of challenges. As the show was designed to be both a tour and a resident show, the projector would always need to be placed in the house, projecting directly towards the scrim. This meant that light from the projector also punched through and was partially illuminating the subjects behind. In addition, the contrast ratio was not as great as hoped, meaning that there was a distinct line between the projected and non-projected portion of the scrim. The director could temporarily hide the performers behind panels covered with duvetyn, but this was just a temporary solution.

The projection team required the cooperation of the lighting designer and the set designer. During rehearsals, drapes were being used to line the performance area. The fabric was black, but not designed to absorb light. Due to this, it allowed the projection to also be viewed by the audience, providing a double image when viewed at certain angles. By replacing the rehearsal fabrics with theatrical, light-absorbing fabric, the first challenge was resolved. The other issue was solved through understanding human eyesight. There was no means of increasing the contrast of the projector, especially with the necessity of how bright the projection needed to be, and the choice of scrim was a necessary component of the effect. In order to change the perceived contrast, the lighting designer bathed the audience in a low-level blue wash. Since the lights were directed towards the audience, it changed their perception, making the projected image appear as the director had hoped.

POWER AND RIGGING

Power and rigging are two aspects of the design where outside departments will need to be consulted and will likely be involved in installation. The projection designer will need to consult with the lighting designer and scenic designer in addition to the appropriate venue personnel in order to determine whether proposed design elements are plausible. This is part of the backward design process, understanding all aspects of how the design will be realized prior to creating the content. If the projection designer does not properly collaborate with the other designers and technical staff, then there is a high likelihood that concessions will need to be made, potentially eliminating much of the projection design.

Large-venue, high-brightness projectors often require power greater than a standard 120V 20A service. Many require greater voltage and may require higher amperage as well. Also, many designs require multiple projectors, making this an even bigger issue. Even projectors that use regularly available power may not have the appropriate power in locations where the projectors are required. It will be important to stress that power provided by a dimming system will not be acceptable (relay power from a dimming system may be tolerated, but there is still the risk of electrical noise impacting the projector). The design team will need to provide the master electrician with the power draw of the displays, especially inrush current (the extra power required at startup), and any additional power needs for accessories.

Assuming that all power requirements are able to be reasonably met, the projection team will also need to consider electromagnetic interference (EMI). EMI impacts the video system through the signal distribution. Depending on the proximity of projectors or other video displays, there is a high likelihood that the signal distribution will need to run near power cables for lighting instruments. Variations in power (especially from a traditional dimming system) can cause a disturbance in the electrical signals in nearby data cables. There are means to minimize this impact, including shielding, distance separation (if on truss, run power lines on one side with data lines on the other), and when they must cross, do so at 90 degrees.

Most projectors, from small home-theater projectors to large-venue projectors, will have some means of being suspended. Even when they do, it is imperative that rigging professionals are involved in their mounting to ensure safety. They will be able to choose the properly rated hardware and a secure mounting option for projectors that do not have rigging points. Any position where the projector is lifted (such as truss towers) technically is rigging, not just suspended loads. The projection team will need to be able to provide the lead rigger with the weight of all projection equipment that needs to be suspended (projectors, video wall, enclosures/support system, cables, projection surface, etc.).

INSTALLATION AND FOCUS

Over the course of this text, we have discussed many factors in how the projection should be designed, from content to display. The design team will have determined factors such as what display technology to use and should have worked with the rest of the design team on how the system should be put together. Very likely concessions will have to have been made along the way and this will make some alterations in the design. We will look at how the most basic system would look and how to deal with them separately.

In the most basic system, as is the case with a projected backdrop only, the design will likely consist of front projection connected to a playback source, likely a computer with a cue-based media server. Even in this, the most basic system, there will be challenges which present themselves in a theatrical setting. Referring back to Figure 8.1, there are essentially three positions for this scenario. The first is back of house, elevated just above the audience. This provides the best coverage of a backdrop, but will be the cause of significant shadows and will be most in the eyes of the performers. This may be a choice for projection on a downstage drop or scrim. The second is still front of house and will either be installed in a followspot position or potentially suspended from a catwalk. This will likely reduce the amount of shadow and will be easier on the eyes of the performers, but will have less of a reach upstage (proscenium and borders will limit the height of the upstage image). Finally, a position on stage, close to the backdrop, will have the least amount of shadow, but the steep angle will add a keystone distortion and may be too close for a large image. Plus, with a traditional wide-angle lens it will be impossible to get the entire image in focus; instead, a more expensive ultra-short throw lens will be required.

BUILDING A SYSTEM   181

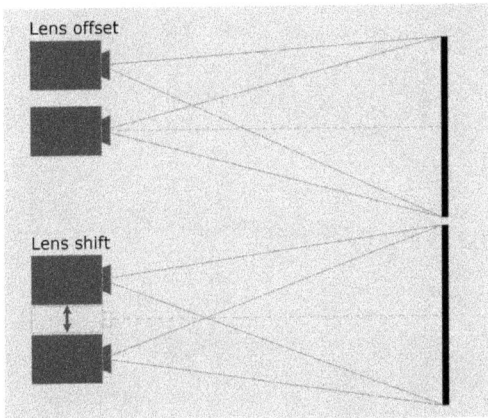

Figure 8.2   Optic design allowing for off-axis projection.

Projectors are intended to be on axis with the projection surface. This means that they should be directly perpendicular to the center of the screen. Some projectors designed for home theater or certain business applications will have an offset in the optics allowing them to be above or below the projection surface, keeping them out of the line of sight of the audience. With large-venue projectors, many will have the option of adjusting the position of the lens in the light path; this is known as lens shift, and sometimes as image offset. This allows for the projector to be outside of the physical center position, yet position the image as needed.

There are a few issues with off-axis projection. The first is when the projector is at an angle that exceeds the optical alignment; the rectangular image becomes trapezoidal. This causes two issues: an obvious size difference from one side to the other which causes the second problem of uneven brightness from one side to the other. The shape of the image can be corrected using digital alignment either in the projector itself (keystone adjustment) or through the software in a process known as corner pinning (reference points of the corners of the display surface used to match the image). This reduces the number of available pixels actually being used to create the image in that area, which can impact the quality of the

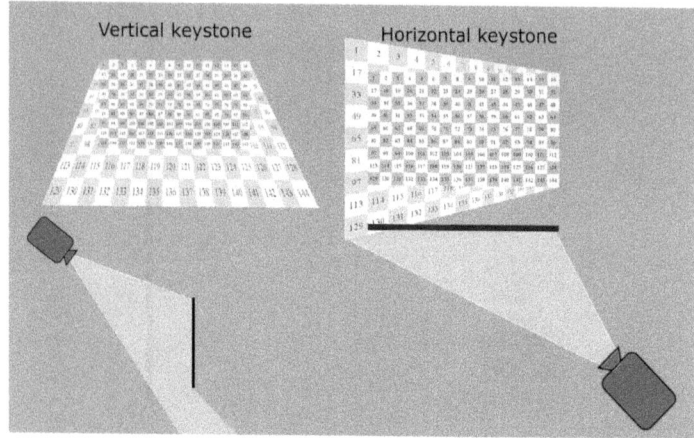

Figure 8.3 Example of digital processing to correct keystone.

image. Generally, the uneven brightness is not adjusted, as that will be a greater level of processing that is not always worth the effort. It may be possible to adjust the uneven brightness within the server software as needed. The third issue, as was mentioned in regard to short throw projection being off-axis, is that it will be impossible to have the entire image in focus. As the center of the image is often the concentration point of the content, when installing, this should be where the image is in focus, allowing the closer and farther edges to fall out of focus. These issues become compounded in installations which require multiple projectors.

LONG-DISTANCE DISTRIBUTION

In order to get signal to the projector, the designer has two basic options: have a media player close enough to the projector that it is within specifications of the method of distribution (may require remote operation of the media player), or use an alternative means of distribution. As mentioned in Chapter 6, an HDMI signal can be converted to SDI, which has a greater transport as long as HDCP is not present for the HDMI signal. If the projector cannot take an SDI signal, it can be converted back to HDMI. Each conversion will add latency to the signal, which could create issues if the visuals require alignment with audio or to match with live action on stage.

Video extenders are a necessity for almost all projection designs. Three methods of extending a video signal are via an Ethernet cable (a couple methods for this), a fiber optic cable, or a wireless signal. The most common method of extending is using an Ethernet cable. The type of protocol used to translate the signal will vary depending on manufacturer and will require a transmitter and receiver, both of which will require power. These signals are not compatible with IP, and thus cannot be introduced to a network switch. If using installed network lines, it is imperative to know that they are "dry" lines (point-to-point connections that do not pass through any network equipment) and the length between points, so as not to go beyond the capabilities of network cable. Some video extenders translate the signal to IP. These can pass through network switches, so long as they are capable of handling the extra bandwidth. It is recommended that these switches not be part of any other network or that they are managed, being able to separate video from other traffic and to provide the necessary resources. There are also a few standardized protocols, such as HDbaseT (https://hdbaset.org/), Software Defined Video over Ethernet (SDVoE, https://sdvoe.org/), and Network Device Interface (NDI, https://ndi.video/). As these are standardized, equipment from different manufacturers may be used together, with some sources and displays being able to use one protocol or the other without a secondary device.

One of the best methods, though often the most expensive, is transport over fiber optic cable. This offers the flexibility of long distances, the potential for multiple signals over a single line, light weight, immunity to EMI, low latency, and extreme reliability. The simplest method of transport is an all-in-one cable. These are powered by the devices which are connected (no additional power required) and can transmit video up to around 200 feet (~60 meters) depending on the bandwidth necessary. They are unidirectional, meaning that the installation crew needs to pay attention as to which end is for the source and which is for the display. The video conversion takes place within the connectors.

Fiber optic video transmission often requires an external transmitter and receiver, similar to signal extension methods using Ethernet cable, where the greatest distance of the signal path occurs over the fiber, but short copper cables are used at either end to connect to the source and display. These systems can reach much greater lengths than all-in-one fiber optic cables, especially for higher bandwidth communication. The system will specify the type of fiber optic cable (single-mode or multi-mode, including OM rating) and connector (ST, SC, LC) required for the system. While these are fairly easy to obtain, this does mean that they usually need to be purchased or rented together to ensure compatibility.

Finally, there is wireless transmission, using radio frequencies rather than wired transmission of data. Traditionally, sending show-critical video as a wireless signal has been discouraged due to issues with reliability and latency. Even with advances in technology, the use of wireless for video may require equipment that is too costly for many applications, but may be possible for some productions. If possible, when wireless transmission is necessary, transmitting commands to a remote player may be more successful than transmitting the video itself. The main reason that wireless is often avoided is that during installation and rehearsals, everything may appear perfect, but when an audience and all of their mobile phones are in the room, they can interrupt wireless transmissions.

There are some factors to consider when looking at wireless. Video quality can suffer at greater distances from transmitter to receiver. The equipment should exceed the requirement as there are additional factors that can impact the video quality. As video signals will require a certain amount of bandwidth (depending on resolution, frame rate, color depth, etc.), these requirements may not be met by all choices. Lower bandwidth requirements will likely be more reliable. The best choice will also have a dynamic frequency selection to avoid interference from other wireless devices. External antennae can offer improved performance and range, with the build quality of the antennae impacting the strength and stability of the signal as a whole. Some additional features may be multicast capability (one transmitter to multiple receivers), ability to transmit control signals, and even on-screen displays.

DESIGNING WITH MULTIPLE PROJECTORS

Many projection designs use multiple projectors instead of just one. This can be for different reasons, including stacking the image for brightness (exact same image on top of one another), blended projection to create a larger image than a single projector can create, or the simple fact that there are multiple projection surfaces that cannot be covered by a single projector. Each of these scenarios may require additional equipment to distribute the signal to the additional projectors.

When the intensity of a single projector is not sufficient for a design, one of the simplest methods of increasing output is to stack projectors. This is easiest to achieve if the projectors also have lens shift capabilities, but can also be achieved with the simplest of projectors. The projectors should be identical if possible, but should at a minimum have the same resolution and frame rate. The signal is replicated, either at the source, or more often through a splitter or a distribution amplifier. A splitter takes a signal and simply replicates it while a distribution amplifier can restructure the signal in order to increase signal range. The images from each projector are aligned to occupy the same area, pixel for pixel, using a focus grid. All light from the projectors is added to increase the brightness. This includes the levels of video black. While this should not really impact the contrast ratio, it will have an impact on blackouts, the apparent edge of the projected area, and can desaturate other colors in the design. Misalignment of the projectors will have the effect of making the image appear out of focus.

Another method which can have an impact on the brightness of an image is using multiple projectors to project a portion of the image. This is often done to increase the area that is covered with an image, such as an extremely wide area. The process overlaps a portion of the images to make it appear seamless. This is known as edge blending. This technique can be seen in Figure 8.4. In this example, the intended projection area is a cinematic aspect ratio

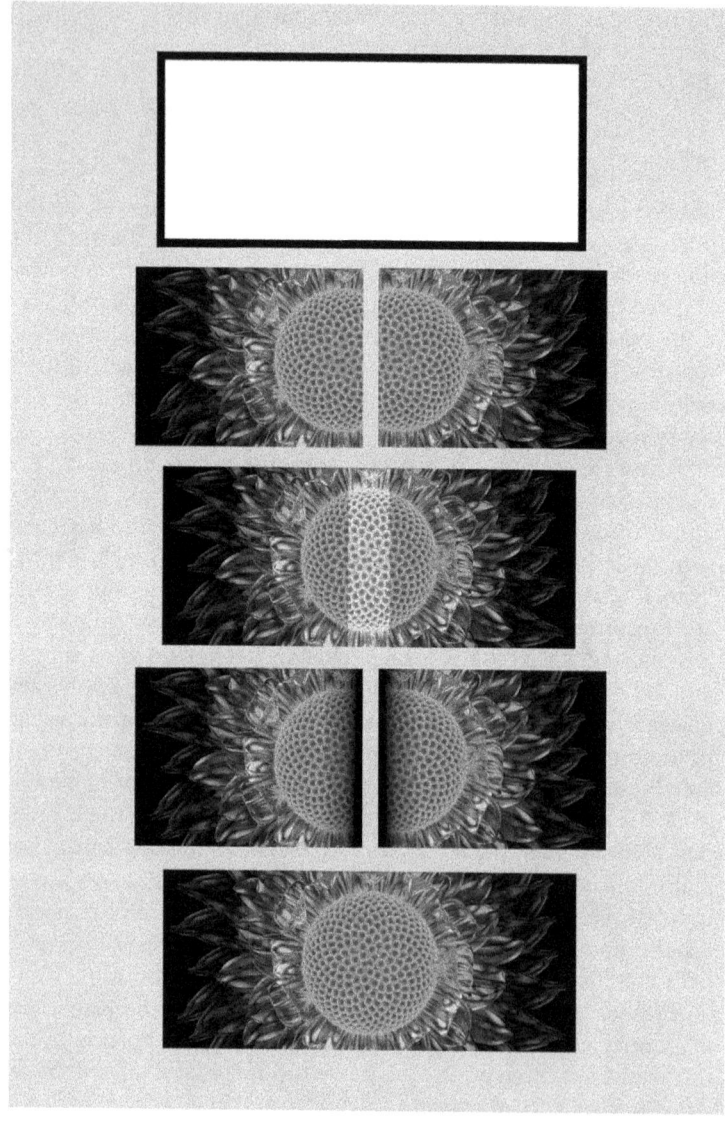

Figure 8.4 The process of edge blending to create a larger image through sending a portion of the content to multiple projectors.

Source: Original image "Sunburst Daisy" by Ryan Bliss (digitalblasphemy.com). Edited by Davin Gaddy. Used with permission.

of 2.35:1. Two 4:3 aspect ratio projectors are blended to create this image. As illustrated, where the two images overlap, the light intensifies. In order to avoid this, the overlapping areas each have a gradient applied. This makes the overlap seamless. This gradient can be applied through the media server or at the projector if equipped to do so. The blending can take place over multiple edges, sides as well as top or bottom edges. Often, when projection systems are designed to illuminate the floor of a sports arena (often seen for professional hockey), there will be six zones stacked and blended. One caveat is that the blend zone is still apparent when the image is mostly (or entirely) black. The only way to avoid seeing the blend zones at that point is to raise the black levels in the non-blended areas. This is an uncommon situation and most often ignored.

The designer will need to know what overlap percentage will be required when creating content. In general, the blend zone will be 10–20 percent of the image. The region of overlap consists of repeated pixels by both projectors. It can be less than 10 percent, which may make the blend area more apparent due to the sharp gradient. It can be greater, making a smoother transition, but at the cost of usable pixels. The designer can use either a predefined projection grid that is covering the span of the resultant image or an individual focus grid with the blend zone established for each projector. Often, large-venue projectors have internal test patterns with a grid pattern that supports edge blending. For complex blends, if budget allows, the designer can employ an automated system using a camera and specialty software to process the blend in mere moments, while it can take considerably longer with a talented projectionist.

With stacked projection it is critical to have matching resolution and frame rates, while edge blending requires even greater compatibility of projectors. With stacked projection, any difference in color reproduction due to differences in projectors (including age or lamp life) is just blended together. This may create some color variation (beyond being more muted due to the brightness change), but will likely not have an impact on the overall image. However, when the projectors are side by side in an edge-blended configuration, the color and intensity differences can be very

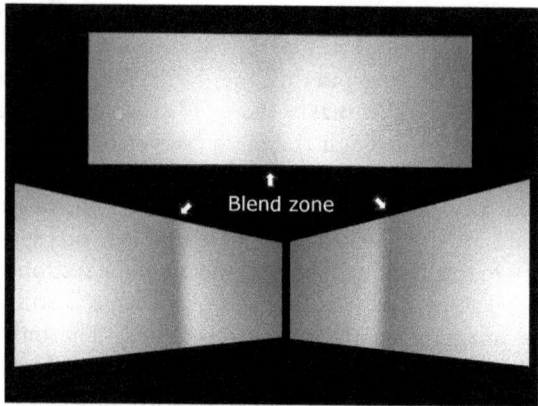

Figure 8.5 Hot spots with edge-blended rear projection: hot spots can ruin an edge blend using rear projection with too great a transmittance, even from different angles. This is exacerbated when using short throw lenses. If space allows, long throw lenses offer a more collimated image, reducing this. If not, the use of a lower transmittance screen will be the only method of properly blending.

noticeable and distracting. If calibration is needed, it may only be available under certain conditions either with specialty software or high-end large-venue projectors.

There are a few other issues which can be present when using multiple displays, including video walls which require more than a single processor to drive all of the panels. The most common issue that occurs is a visual artifact, caused by a timing error, known as tearing, where there is a horizontal shift between lines of the image. While this can occur within a single display due to non-matching refresh rates, this is more common between displays creating parts of the same image. The tearing is most noticeable in horizontally moving visuals. This can be corrected through vertical synchronization, but more commonly through the use of generator locking (commonly known as genlock), which uses a reference signal to synchronize imaging sources. The use of a black burst (a video signal with no picture data used in legacy equipment) or tri-level (higher frequency for modern equipment) signal enables a precise frequency of when frames start, synchronizing the timing plane for the entire connected system. An improperly timed genlock signal could introduce judder in the video signal. In addition, a frame-synchronized system adds at least one frame of latency. Also, not all gear has the ability to receive genlock.

DOCUMENTATION

For every design there will be required documentation. This will include everything from pre-production design elements, to the cue list, and what it takes to preserve the show. As projection design is the relative newcomer, this type of documentation is not as regimented as in other disciplines. Instead, a lot of documentation borrows from other industries. Some of the paperwork has been discussed in detail, such as the design elements for content. This included the overview of the design, storyboards or similar pre-production ideas, as well as previs elements midway through. The designer should have an order of operation, what has to happen and in what order during the production process.

The next bit to record are the technical specifications and installation requirements. These documents will need to include some specific details in order to enact a successful design. The first document will be a comprehensive list of all equipment needed for the projection setup: projectors, projector lens (if available to exchange), media server and other content playback devices, distribution requirements, and all mounting hardware. When listing equipment, it will need to be as specific as possible, such as the detailed information on the projectors (brightness, resolution, signal connectivity, and power). If there are special projection surfaces or treatments for set pieces or any other special considerations, this will be the place to make these specifications.

Installation documentation will be two-fold, the ground plan and elevation to show where display elements will be within the venue and possibly some of the placement of distribution cables, while the schematics of the rest of the system design will be separate documentation, including the totality of signal flow. The designer may collaborate with the lighting designer to include projection elements in the light plot if they occupy the same space so as to help in placement. Any special considerations for rigging or mounting should be included here. This documentation should include power requirements as well as load calculations for rigging. Additional documentation will include a list of network information. Specifically, it should include all IP addresses (if fixed), the name of the device (if available), the media access control (MAC) address (a unique hardware identification for a device), user name and password as needed, as well as specific comments regarding its purpose (especially for primary and backup systems). If it will be a touring production, certain elements that are required will need to be specified to ensure that every audience is able to experience the show in the same manner, or as reasonably close to the original as possible.

All content will need documentation. This will include the guidelines for creating and formatting new content (assuming a long-running production which may need new content to stay fresh), including the resolution, file formats, etc. In addition, backup files for all current content must be maintained, including elements used in its creation and any attributions required. There are several naming conventions for the files, but each should have identification as to what is in the file, where it is to be used (if needed), and some means of identifying the latest version. The files should be stored both locally and remotely if the show will be running more than a few weeks.

As with all departments, cue sheets will need to be created. This will include all timing and descriptions of how they will be executed, be that a called cue or automated through show control. It is recommended that operational instructions be included in the cue sheets as well, including startup and shutdown procedures. All designs should include some form of emergency protocols for when show elements do not work correctly. These elements need to be documented in the cue sheets as well, most likely as an appendix. In order for the operator to actively troubleshoot issues that may arise, documentation on how the system integrates within its own boundaries and including other show control systems is necessary.

Even with limited-run productions, the system will need to be maintained. The show documentation should include calibration procedures for all displays. All projection should include photographs of where the image is to be produced so that they can be corrected if misaligned. This may require specific test patterns and photographs of those being used in place. It should be noted if there are specific points where the projectors need to be focused, especially when using a long throw lens which may have a deeper focal plane. Any procedures for correcting common issues, such as signal flow issues which may be caused due to handshake failures or other hardware failures, need to be included.

Some additional information that may be documented for long-running productions may not be necessary for those running a month or less. These may be kept on hand for venues with installed systems as well. Information regarding safety procedures, such as projector lamp changes which could pose personnel risks, should be included. If there are any compliance certifications required (specifically for laser projectors), these should be maintained with other design documentation. Finally, vendor and supplier contacts for equipment and materials, especially if there are warranty and technical support contacts, should be included. By preparing thorough and detailed documentation, all aspects of the design, setup, and operational procedures are clearly communicated.

SPECIFICATIONS

In order for the designer to specify the equipment needed for an optimal presentation, the designer needs to be able to calculate what the design will look like. This has been discussed in various chapters, but now we will discuss some of the particulars, including the math. When the particulars are known, it can come down to preference and budget to finalize the design. Some things to figure out: image size, brightness, pixel size and density, and overlap needed for circumstances such as blends/mapping/masking. Special circumstances such as interactive elements will likely need to be dealt with on a case-by-case basis and is beyond what can be discussed here. So, we will use some examples to introduce how this might need to be determined.

To start, let us assume that the director is looking simply to fill the upstage cyc with a background image. Let us assume that the director is looking to replace the 40 x 24-foot drops that are normally rented. The first thing to consider is the size of the image. Projectors typically come in just a few standard aspect ratios: 16:9, 16:10, 4:3, and rarely in modern projectors, 5:4. In order to figure out which type of projector can come closest, the designer will change a ratio to a decimal (16:9 is the same as 16 / 9 = 1.7777). To achieve the width, divide by the ratio to get the actual height (40 / 1.7777 = 22.5) which falls short of the height necessary. By multiplying the height of the desired image by the ratio, then the width can be determined (24 ★ 1.7777 = 42.6648). This means that to use the full height of the projected image, the designer exceeds the width. To achieve the height, the designer will need to overshoot the sides of the cyc, blanking the portion of the image that would exceed the projection surface (this may be acceptable with proper legs to absorb excess light). The designer would need to make a choice, either to keep the width and be too short or overshoot the width in order to achieve the height. Increasing the size of the image does mean that the brightness will marginally decrease. A different projector with a different aspect ratio could also manage a little better (a 16:10 projector would have a height of 25 at the width of 40).

Let us next assume that the designer has a 10,000-lumen projector available to create the image on the cyc. To see if this is a good start for brightness, the designer will need to do some calculations (refer to formulas found in Chapter 6). If the projector available has a 16:9 aspect ratio, and the designer has chosen full height, thus exceeding the width, then the image is 42.7 x 24, with a square footage of 1,024.8. So the amount of light falling on the surface is 9.8 lumens per square foot (foot-lamberts) if the cyc had a gain of 1.0 (more likely it will have a gain closer to 0.5, reducing to 4.9 fL being reflected). This falls dramatically short of the target goal of 70 lumens per square foot projected (to contend with stage lighting and low contrast due to ambient light), even dramatically short of what is used in a dark movie theater.

Without changing any equipment, the designer can instead look to see what size image is achievable with the target goal for

Figure 8.6 Illustrating mismatched aspect ratios: when the aspect of the projection surface does not match the aspect of the projector, the designer will face the choice of seeing the entire image but not filling the surface, or filling the surface but losing some of the image due to overspill.

brightness. Let us say that the content will be brighter and not needing a lot of contrast, so the target goal is 50 lumens instead of 70. This means that the same projector on the cyc would be able to produce a 100-square-foot image (10,000 * 0.5 / 50 = 100). This means that the image size would be around 13.25 x 7.5 to get the brightness desired. Either that, or figure that additional or brighter projectors will be needed (50 * 1,024.8 / 0.5 = 102,480), increasing the need ten-fold. If the cyc had a higher gain calculated, for instance at a 0.9, then the design would still need (50 * 1024.8 / 0.9 = 56,933) close to a 60k projector to achieve desired results. This is where the designer needs to determine expectations for how the design will look. Should the image be darker than desired, smaller than desired, or some compromise of the darker and smaller? How will the content look being too dark or washed out? Can the designer budget for brighter projectors? Will there be sufficient power for brighter or more projectors? The designer will also need to consider that the recommended brightness is for replicating the initial drop, with the same vibrancy of colors. A less bright image with more muted colors can look perfectly fine as the audience will not be doing a side-by-side comparison with the original.

At least with this size of image, the audience is unlikely to be close enough to discern individual pixels, assuming a stage depth of around 35 feet. For simplicity, let us assume that there is very little screen door effect and that each pixel nearly connects to its neighbor. An image that is 42.7 x 25 will have individual pixels that are roughly 3.5 inches square. This means that anyone closer than 70 feet with standard 20/20 vision might be able to distinguish individual pixels, depending on the content (image and motion). Visual acuity is different between a projected image and a direct-view display, as projected pixels have less space between them. Direct-view displays will have variances between the pixel density as well as the pixel size, both of which need to be taken into account.

BOX 8.5 UNDERSTANDING 4K IMAGE CREATION

Some projectors use a technology known as pixel shift in order to increase the apparent resolution of the image. In particular, this is used in a projector which uses an HD imaging device to produce a 4K image. As the 4K image is not being developed as native resolution, some will call this a "Faux K" image, implying that it is false.

This is different from just scaling an image, but is more akin to a progressive compared to an interlaced image. The resolution of the content is the same; there is just a difference in how the display recreates that image. Most commonly, the projector will have an HD chipset. As 4K is a doubling of height and width of an HD image, the projector has an additional component, known as an optical actuator, to the optical path which allows the shift of where the pixel is displayed. This same technology is also used in some cameras in order to increase their resolution.

To produce the image, the optical actuator shifts the placement of the pixels one half-pixel diagonally, which creates an overlap of the pixels with the center being in the midpoint between the corners of four pixels. Unlike interlaced video where each field has alternating lines of video and blank lines, each "field" of a pixel-shifted image is a complete frame. Some viewers state that this makes the image less sharp. Due to the distances that most audiences would be viewing this from in entertainment venues, it is unlikely that even a trained eye would notice. As technology advances, this will likely continue to be almost invisible to the human eye. Where it makes a difference to the designer is in cost (pixel shift costs less than an equal higher native resolution).

Higher pixel density will offer a crisper image and allow for finer detail. This allows the viewer to be closer to the image as well. The surface chosen for the projection as well as the angle of the viewer can all be a factor in the quality of the image. The human eye can distinguish details of about 60 pixels per degree of viewing (known as 1 arcminute). If detail is desired in the image, a minimum of 4K resolution is recommended to maintain clarity. That being said, much of the projection design for live entertainment does not need such clarity.

PROJECTION DESIGN AS PROJECT MANAGEMENT

Projection design can easily fit within the science, technology, engineering, art, and math (STEAM) fundamentals. This sometimes means that the designer needs to put together the optimal team as would a project manager. The ideal means of delivering all components in a timely fashion requires more than just creative vision. A designer should not feel stifled by using this method, but use it to allow freedom through group think.

For those who may not be familiar with tenets of project management, there are four key areas. The project will strive for better control with details in budget and schedule. It will succeed through better communication, which can minimize mistakes. It makes for better customer relations, which in this case is primarily internal customers (the rest of the production team), but also has an impact on the end product for the audience. The first three key areas highlight the final area, which are higher-quality results.

The design will have dealt with the scope of the project, how the director envisioned video to fit within their vision of the production. The designer needs to be precise in what is deliverable, being sure to avoid vagueness. The project will adhere to a schedule. This schedule will really depend on the scope as there are so many variables that go into the design. They need to be kept realistic. For instance, experimental interactive elements may need development time, or if filming of actors in costume is necessary (will there be enough time for the costume designer to have the costume built in addition to filming and editing?). As projection is often outside of what is regularly an element equipped by a venue or production company, a large portion of the budget may be allocated to rentals. This may be challenging to the design team as they may not have access to the equipment in order to test how things will look or work together. This adds a stressful challenge that may make designers take fewer creative risks. After all, the quality of the end result encapsulates the entire design process. After the production opens, always hold a postmortem (an analysis of an event to determine what was successful and what was not).

Some final thoughts: as has been stated many times, the process takes a lot of back and forth. The end is the beginning and the beginning the end. Projection design can take on a life of its own, encompass a great amount of space, and be a part of almost everything we see in a production (lights, set, costume, props, and makeup). It has structure, as it is bound by the laws of physics, but it is fluid in that human vision is fallible. The projection design is continually evolving, meaning that there is no one "right" way to accomplish it, but there are certainly means of more effective designs. What is standard practice today may seem archaic in just a few short years. It is as ephemeral as the projection itself. This gives the designer the promise of something new with every production.

BOX 8.6 SUMMARY

- Information gathering at the venue is crucial to system design.
- Tools for a site survey include a notebook, a camera, measuring devices, a light meter, and a flashlight.
- The designer will be choosing between front or rear projection, each with their own space requirements.
- Properties of a projection surface include gain, viewing angle, color, and texture.
- Non-traditional projection surfaces will have unknown properties and may require periods of experimentation during the design process.
- Projection installation will require the assistance of the master electrician and potentially the head rigger.
- Long-distance signal distribution can be carried over Ethernet cables (some can be used in a network), fiber optics, and wireless transmission.
- Multiple projectors can be used for brightness (stacked image) or cover a larger space (edge-blended).
- Every design requires proper documentation. This includes design specifications, cue sheet, and archival information.
- The designer may need to make concessions after doing the math for size and brightness of the projected image.
- While projection design is creative in nature, the design process should mimic project management to deliver quality results.

GLOSSARY

Acoustic transparence The maintenance of acoustic fidelity as sound passes through a material, such as a projection screen.

Aliasing The jagged edges along the edge of objects created through inadequate sampling techniques in a digital image.

Alpha channel The transparent channel of a digital image file; this allows for easily layering images on top of one another.

Ambient light Any light on a display surface originating from sources other than the intended light of the projector, which greatly reduces the contrast ratio of the image.

Analog Video signal containing color and luminance information using voltage regulation.

ANSI American National Standards Institute: a private, non-profit organization that administers and coordinates voluntary standards and conformity.

Aperture The device which controls the amount of light entering or leaving the optical system.

Artifact An unwanted distortion or defect in a video signal, often stemming from poor-quality compression or transmission.

Aspect ratio The numeric comparison of the size and shape of an image with regard to the width to height. May be represented in two whole numbers or as a fractional number compared to one (16:9 or 1.77:1).

Bandwidth 1) The frequency range within a band of wavelengths in the electromagnetic spectrum, used for transmission of data. 2) The measurement of the capacity of data transfer in an electronic communications system.

Bezel The frame around a video monitor which acts as a protective covering for the edge of the transparent cover as well as other non-display components of the device.

Bezier curve A parametric curve used by a computer to smooth a line, especially when mapping onto a spherical or curved shape.

Bitrate The rate at which bits are transmitted along a digital network, measured in bits per second (bps). The measurement of how much data is transmitted in a given amount of time.

Bitrate, constant A consistent bitrate used through playback (decoding), allowing for fast loading times.

Bitrate, variable A dynamic bitrate that changes depending on the level of detail required in the decoding process. More resource-heavy, but provides higher-quality video while maintaining smaller file sizes.

Black burst A video signal with no picture data (full black image), used as a reference signal to synchronize frames across devices.

Black level The level of brightness at the darkest part of a display which has no color value (black).

Blanking 1) In interlaced video, this is where black (no image) is used to switch between fields (vertical as well as between scan lines (horizontal)). 2) It is the process used to adjust aspect ratio by eliminating image data to horizontal or vertical lines at the edge of the image.

BNC Bayonet Neill-Concelman: A locking connector for coaxial cables.

Calibration   The adjustment of a display to accepted standards in order to present an image in as close a representation of the source material as possible.

Camera   1) A device for capturing images and communicating those images. 2) A point-of-view marker used in imaging software.

Chromatic aberration   An image distortion from a failure of the lens to focus all colors to the same point, which creates an outline of unwanted color along the edges of objects.

Chrominance   The color component of a video signal.

Codec   An algorithm used by computer software which allows for the compression and decompression of a video or audio file, allowing for the reduction of file size.

Color bars   A video test pattern which allows the video engineer to calibrate the video display, utilizing reference values of chrominance and luminance.

Color gamut   The range and depth of a color available by a given display through additive color-mixing of red, green, and blue.

Color resolution   The total number of colors available in a given video image, expressed in bits per pixel.

Color saturation   The brilliance and intensity of a color in an image. The greater the saturation, the purer the color appears.

Color space   The graphical relationship of color based on hue, saturation, and luminance.

Color temperature   Measured in degrees Kelvin, this represents the value of white light in how it is represented in color purity. A lower temperature will be shifted more towards the red end of the spectrum while a higher temperature will shift more towards the blue.

Color wheel   1) A graphic representation of the relationship of colors. 2) Component in a single-chip projector which allows for a range of colors to be displayed.

Compositing The process of combining digital visual resources into a singular output.

Compression A method of reducing the file size in order to either save storage space or to reduce bandwidth in transmission.

Container A metadata file format which packages various parts of the video data.

Content Digital files containing video or still images with the intent of being displayed to an audience.

Contrast The difference between light and dark.

Contrast ratio The measured comparison between the brightest white to the darkest black that a display can provide. Measurement can be given as a full bright image to completely off, or the ANSI method comparing 16 alternating white and black rectangles in the same image.

Convergence The alignment of video elements within an image to create the whole.

Corner pin A tool used to manipulate a video signal to match the four corners of a physical or virtual display surface.

Depth of field The distance between the nearest and furthest an object can be in focus in a photographic image.

Depth of focus The range in which a projected image appears to be in focus in a given plane.

Digital Information processed by computers by means of binary code.

Dispersion of light The change in angle of refracted light resulting in the separation of visible light into its constituent colors.

DisplayPort A digital display interface capable of carrying information for multiple displays, audio signals, data, network information, and power.

Distortion, barrel An optical deviation of rectilinear projection to where straight lines bend outward from the center.

Distortion, pincushion An optical deviation of rectilinear projection to where straight lines bend inward towards the center.

Distribution amplifier An electronic device used to duplicate and transmit a single video signal to multiple devices while maintaining signal integrity.

Dithering The process of making digital images appear smoother by adding random noise during the digitization process.

Douser A physical barrier between the light source of a projector and the display surface. This can be an internal component, being positioned prior to light being emitted through the lens, or an external accessory placed in front of the lens.

DVI Digital Visual Interface: A digital video interface standard for transmission of uncompressed content.

Dynamic iris A mechanism in select projectors which is integrated between the projector light source and the lens, opening or closing to adjust the light output in order to enhance contrast performance.

Dynamic range The range of brightness that can be expressed in a digital image.

Edge blending A technique used to visually combine overlapping sections of projected images into a single seamless image.

Ethernet A physical and data link networking protocol for connectivity of computers and other devices to a network.

Extender A colloquial term describing a device in which the range of digital signal is increased beyond traditional cabling methods.

F-stop The aperture measurements in an optical system which controls the amount of light that passes through the lens.

Field A portion of a still image, either even or odd lines, which is displayed sequentially to create the impression of motion in a video. Two fields are interlaced to comprise one video frame.

Focal length   The optical distance from the point where light from a projector can focus.

Focal plane   The distance where a projected image appears in sharp focus which may be limited (narrow) or a given range (deep).

Foot-candle   A unit measuring illumination which is one lumen per square foot.

Foot-lambert   A unit measuring luminance of a surface which is one lumen per square foot.

Frame   A full image which is displayed in sequence to create the impression of motion.

Frame rate   The amount of times that a frame of video is refreshed per second (fps), measured in hertz (Hz).

Front projection   A reflective form of projection, where the projector is on the same side of the projection surface as the audience.

Gain   1) An amplification of the electrical signal sent from the image sensor, which increases brightness of the image, but reduces quality. 2) The measurement of the reflectivity of a projection surface.

Gamma correction   An adjustment which impacts brightness and contrast to make sure that the output signal appears the same as the captured image.

Genlock   Generator locking: a common reference signal used among devices to synchronize sources and displays.

Geometry, projector   How a projected image appears based on the relation of the projector to the projection surface.

Graphics card   A printed circuit board that generates digital images in a computer system.

Grayscale   A monochrome image which only utilizes luminance values. Test patterns in grayscale allow for proper brightness and contrast settings in a display.

Half gain The angle horizontally off perpendicular at which the intensity measurement of the image has declined 50 percent from peak brightness.

HDD Hard disk drive: an internal non-volatile digital data storage medium which comprises a spinning magnetic disk, paired with a magnetic head on a moving actuator arm.

HDMI High-Definition Multimedia Interface: a proprietary digital interface standard for transmitting uncompressed video and either compressed or audio data.

Hertz The unit of frequency equal to cycles per second, abbreviated as Hz.

High dynamic range An increased color gamut in an image due to expansion in luminosity, providing a greater contrast and detail than standard dynamic range.

Illuminance The luminous flux incident on a unit area.

Interlaced A video process of displaying two fields (odd lines succeeded by even lines) of an image to create a frame of video.

Jitter An irregular or inconsistent frame rate due to poor signal transmission.

Judder A perceived uneven or jerky appearance of movement due to frame rate inconsistencies.

Keystone The visual distortion of a projected image due to the projector not being perpendicular to the display surface. Resultant image is no longer rectangular, but trapezoidal.

Keystone correction An electronic correction of an optical misalignment in order to create a rectangular image, resulting in a reduction of the quality of the projected image.

LAN local area network: an interconnection of computers and devices within a limited area.

Latency The delay between the image being processed and displayed.

Layer A component of video where a level of aspects (still or moving images, audio, text, and effects) can be independently manipulated, giving hierarchical priority when rendering.

LED light-emitting diode: a semiconductor device that converts electrical energy into light.

LED display An array of LEDs utilized as pixels in a direct view display for digital images and video.

Lens shift A process of mechanically moving the lens in projectors equipped to do so which results in the image being correctly aligned with a display area when the projector is not aligned with the center of the surface.

Lens, fixed A type of lens which has a static focal plane. Image size is dependent on placement of the projector.

Lens, long throw A projector lens which has a throw ratio greater than 4, allowing the projector to be at a greater distance from the surface.

Lens, parfocal A type of zoom lens which is able to maintain focus while changing focal length.

Lens, short throw A projector lens which has a throw ratio less than 1, allowing the projector to be close to the surface and produce a large image.

Lens, varifocal A type of zoom lens which does not maintain focus when changing focal length, mainly designed for still image cameras.

Lens, zoom A type of lens which has a range of focal planes. The image size can be adjusted without requiring the placement of the projector to be changed.

Lenticular 1) An array of lenses which alter what can be viewed when the viewing angle changes, giving the illusion of depth when used for 3D images or can be used for a change in a 2D image. 2) A projection screen with a texture of ridges which can be used to focus more of the light toward the viewer (increasing apparent brightness) or to aid in rejecting ambient light.

Light meter, incident A device which reads the intensity of light falling on an object.

Light meter, reflective A device which reads the intensity of light reflecting off an object.

Lumen An SI unit of light measurement used in the definition of intensity output of a projector; specifically a measurement of luminous flux, equal to the amount of light emitted per second of one steradian from a point source having a uniformity of one candela.

Luminance 1) The component of a video signal which carries brightness information. 2) The photometric measurement of intensity of light emitted from a surface per unit area in a given direction.

Lux A unit measuring illuminance, equal to one lumen per square meter.

Masking The process of adding a layer to blank areas of an image not intended to be seen.

Media server A software application or computer system designed to store and share multimedia files. In live entertainment, this can also alter the video image as needed and operate under show control or operator-specific cues.

Mesh warp The digital alteration of an image in specific areas without affecting the entire image.

Mirroring Displaying duplicate content from a computer onto an additional display.

Moiré effect A visual illusion which presents as patterns which interfere with one another due to a resolution difference between content and display.

MIDI Musical Instrument Digital Interface: A communication protocol and digital interface to interconnect electronic musical instruments, computers, and related devices.

Native resolution The hardware-determined resolution of a display, measuring the pixels horizontal and vertical.

Network A method of interconnecting computers in order to share resources.

Nit A measurement of light which is equal to one candela per square meter (from Latin, meaning to shine).

Opacity A measure of the transparency of an image, including the image's alpha channel.

Overlay The process of superimposing graphics or text onto a video or still image.

Overscan When an image is greater than the viewing area of a display, causing the edges to not be processed for viewing.

Persistence of vision An optical illusion that occurs when visual perception continues after the rays of light have ceased to enter the eye. The necessary visual component of motion pictures when a sequence of still images appears as movement.

Pixel The smallest addressable picture element in a raster image.

Pixel density The number of pixels within a given physical measurement (inch or centimeter), but does not refer to the size of the pixel itself nor the resolution of the display.

Pixel shift A method of increasing a produced resolution by altering the position of the image through the use of an optical actuator.

Progressive scan A format for displaying moving images where all lines of each frame are drawn in sequence, as opposed to the two fields in interlaced scanning.

Projection axis The positioning of a projector from the center of the lens to the center of the projection surface. On-axis refers to a perpendicular alignment while off-axis is an alignment other than perpendicular.

Projection mapping The use of software to alter an image with the purpose of projecting visual effects onto a 3D object. The term succeeds spatial augmented reality.

Projector An optical device which produces an image onto a surface through the use of directed light.

Projector offset A value given to determine the placement of a projector in regard to the height of a screen. Lens offset = (screen height + offset height) / screen height.

QoS Quality of service: the measure of network system performance with considerations of availability, bandwidth, latency, and reliability by prioritizing network traffic to ensure a required level of service.

RAM Random access memory: the temporary memory where computer instructions are executed and data is processed.

Raster graphics A file that is composed of a grid of individual pixels that collectively build an image in the form of a bitmap.

Real-time A description of system response where processing happens as fast as required per application in order to respond without noticeable delay.

Rear projection A transmissive form of projection, where the projected light passes through the projection surface to the audience who is on the opposite side.

Refresh rate The frequency of times per second that all pixels in a display are re-energized in the creation of an image.

Render 1) A graphic representation of a particular view of a 3D object. 2) To convert coded content into a usable format for display, particularly from an image or video-editing software.

Resolution The matrix describing the horizontal number of pixels and the number of lines which define an image or display area.

RGB Red Green Blue: the three components of a color display used as additive mixing in a trichromatic fashion.

RJ45 Registered jack 45: a common type of 8P8C (8 position, 8 contact) networking interface for a cable of twisted wire pairs which allows insertion in one orientation with a compatible socket.

Saturation The perceived brilliance and intensity of a color in an image as judged in proportion to its brightness. Greater saturation is considered to be pure while low saturation is considered to be muted.

Scaler A device or software which resizes an input signal to a different resolution, generally to match a display's resolution.

Screen 1) The display area of a video monitor. 2) An optimized surface manufactured with the purpose of receiving a projected image.

Screen door effect A grid-like anomaly in a projected image where a visible space between pixels is noticeable, resembling that of looking through a screen mesh.

SDI Serial digital interface: a family of digital video protocols standardized by SMPTE in order to transport broadcast-grade video.

SI International System of Units: a widely used system of measurement coordinated by the International Bureau of Weights and Measures, commonly known as the metric system (from the French *Système international d'unités*).

Signal flow The connections for which audio and video data is transmitted through distribution to the display.

SMPTE Society of Motion Picture and Television Engineers: a professional society which prepares standards and documentation for video production.

Solid-state memory An electronic, non-volatile storage medium with no moving parts.

Splitter A passive electronic device that sends a duplicate signal to multiple receivers.

Surface The physical component of a projected image where the projected light is viewed.

TCP Transmission Control Protocol: a network communications protocol for the exchange of data packets that ensures data is delivered to the correct destination.

Test pattern An image file with patterns and colors used to calibrate a digital image.

Throw distance The distance required between a projector and the surface to create an image of a given size.

Throw ratio The relationship of the distance from the projector to the projection surface and the width of the image.

Timecode A sequential numerical coded signal used for synchronization, containing data in the temporal units of hours, minutes, seconds, and frames.

Transcoding The process of decoding a video file from one format and encoding it to another.

UDP User Datagram Protocol: a network communications protocol which prioritizes timeliness over accuracy of packet delivery.

Upscaling The process of increasing the resolution of an image, which may result in a poor-quality representation of that image.

UV mapping The process of defining how to represent a 2D image on a 3D object.

Vector graphics An image file which uses mathematical equations to form lines and curves with fixed points on a grid to define shape, border, and fill color.

VESA Video Equipment Standards Association: an international non-profit corporation which supports and sets industry-wide interface standards for personal computing and consumer electronics.

VGA Video Graphics Array: a catch-all term for analog video distributed through a D-subminiature 15 connector. The standard resolution established by VESA is 640x480.

Video black The absence of generating any color in a video signal. A video display still emits some amount of light.

Video signal The digital transmission of video content.

Viewing angle The angle diverging from perpendicular to the center of a screen at which point half gain is achieved.

VRAM Video random access memory: dedicated computer memory used to store graphics data with the purpose of ensuring a smooth execution of the image to the display.

WAN Wide area network: an interconnection of computers and devices over a broad geographical area, generally connecting multiple smaller networks.

Zoom, digital The simulation of a longer focal length lens through the digital manipulation of an image.

Zoom, optical The process of changing the focal length by physically adjusting the relationship of lenses within a lens assembly.

INDEX

Note: Figures are shown in *italics* and table in **bold** type.

4K image creation 194–5

absorption 24, 40
acquisition, of digital content 107–9
action charts 48, 54–6, 62, 64, 88–9
acuity: spatial 33; spectral 33–4; temporal 33
Advanced Television Systems Committee (ATSC) 111
aesthetics 6–7, 98, 113, 123, 169
AI (artificial intelligence), in design process 53
alpha channel 117
ambient light 37, 38, 39, 125, 128, 130, 148, 172, 174, 175, 192; on contrast ratio 130–1
ambigrams 104, 105
American National Standards Institute (ANSI) 92, 127, 128, 129, 134, 153, 160–1, 162

American Standard Code for Information Interchange (ASCII) 17–18, 103–4
amplitude 20, 22, 40
analog 2, 5, 15, 16, 135, 137, 165
animation 50, 61, 108, 131
ANSI (American National Standards Institute) 92, 127, 128, 129, 134, 153, 160–1, 162
aperture 132–3
AR (augmented reality) 98
artifacts 91, 137, 138, 139, 175, 188
artificial intelligence (AI), in design process 53
Art-Net 153, 159–60, 161, 162, 166
ASCII (American Standard Code for Information Interchange) 17–18, 103–4
aspect ratios 94, 109–12, **110**, 116, 119, 123, 185–7, 192; mismatched *193*

asymmetrical arrangement 81
atmospherics 25–6, 75, 130, 177
ATSC (Advanced Television Systems Committee) 111
audio: design 16; embedded 118–19
Audiovisual and Integrated Experience Association (AVIXA) 30, 129
augmented reality (AR) 98
AVIXA (Audiovisual and Integrated Experience Association) 30, 129

backdrops 6, 8, 9, 35, *76*, 76, 83, 93–5, 171, *172*, 180; projected 35, 94, 180; scenic 93, 94
backward design process 178–9
balance 79, 81
bandwidth 114, 149, 157, 160, 183, 184; cables 137; DisplayPort 138; High-Definition Multimedia Interface (HDMI) 136; MIDI Show Control (MSC) 155–6; serial digital interface (SDI) 140
Bartleson-Breneman effect 71, *72*, 131
best practice 2, 32, 38, 78, 169–70
B-frame 115
biomechanics 26–32, *28*
"bird's-eye view" 84, 95
bitrate 116, 117–18, 139, 140
black burst 188–9
black levels 40, 187
blanking 111, 192
blob detection 62
brain 26–7, 28, 29–30
brainstorming 42, 48
brightness 22, 31, 33, 39–40, 128, 131, 191, 197; front projection 172; LED wall 173; multiple projectors 185; off-axis projection 181, *182*; rear projection 171

broadcast standards 140
building a system 167–97, *172*, *181*, *182*, *186*, *188*, *193*

cable length 137, 149, 153, 155
cable pathway 169
cable standards 138–9
calibration 33, 128, 188, 191
camera obscura 29
CCT (correlated color temperature) 20
central processing unit (CPU) 144, 145
CFF (critical fusion frequency) 33
characters: in a story 44, 45, 53, 95, 96, 100, 101; in text 90–1, 104
choreographed video 59
chromaticity 30–1, 121
chromoluminarism 33, 101
CIE (La Commission internationale de l'éclairage) 30–1
cinematic projection 12–13
cinématographe 14
closer objects 77, 85
codecs 113–14, 115, 118, 119, 135, 145
collaboration 46–58
color 68; in content 100–3, *101*; primary 30, 69; secondary 69
color depth 116–17, 139
color harmony 70–1
color temperature 20, 22
color theories 70, 101
color wheel 69, 70, 71
Commission internationale de l'éclairage, La (CIE) 30–1
common experience, in seeing the world 32–4
common language 17
common misconceptions 34–7
communication 8, 16–18, 41, 57, 64, 152; wired 157, 159
compositing 114, 146

INDEX 213

compositional elements 64–79, *65*, *67*, *71*, *72*, *74*, *76*
compositional principles 79–87, *80*, *86*
compression 113, 114, 115, 116, 139; inter-frame 114; intra-frame 114
concave lens 24, *25*
concrete poetry 104
cones 27–8, 32, 33–4, 40
connections 19, 135, 155, 156, 183
consistency 50, 79, 80, 150
containers 113–16
content: color in 100–3, *101*; creation of 10, 35, 116–17, 118, 144, 146; digital 35, 98, 107; royalty-free 107; stock 107–8; text in 103–5
contrast ratio 17, 31–2, 38–9, 40, 129–30, 149; and ambient light 130–1; and front/rear projection 172; and video black 185
control systems 10, 11, 59, 62; *see also* show control, systems
convex lens 24, *25*
cooperation 6, 7, 97, 178
copyright 105–7
cornea 27, *28*, 28, 40
correlated color temperature (CCT) 20
CPU (central processing unit) 144, 145
Creative Commons 108
critical fusion frequency (CFF) 33
cue sheets 170, 190, 197
cues 36, 58, 59, 60, 146, 150–1, 152, 156, 164, 170

depth of field 62, 132, 133
depth of focus 133, 134, 171
design elements 41, 43, 52, 58, 77, 81, 91, 178–9, 189
design process 1, 29, 32, 64, 112, 120, 168, 197; artificial intelligence (AI) in 53–4; backward 178–9; *see also* compositional elements; compositional principles; design elements; workflow
diffuse reflection 23, *25*
digital content 35, 98, 107
digital displays 33, 92, 120
digital media player 16, 141–2, 149
Digital Multiplex (DMX) protocol 153–4, 155, 159, 160, 161, 166
digital storyboard 51
director 7–8
direct-view displays 31, 39–40, 101, 120–1, 122, 148, 194
dispersion, of light 24, 25, 40, 128
DisplayPort 137–9, 140, 149
displays: direct-view 31, 39–40, 101, 120–1, 122, 148, 194; multiple 5, 7, 10, 33, 122, 135, 138, 146, 188; video 17, 98, 179
distortion 24, 77, 134, 172, *172*, 175, 180
distribution 169, 171, 172–3, 179, 185, 189–90; long-distance 182–5
DMX (Digital Multiplex) protocol 153–4, 155, 159, 160, 161, 166
documentation 189–91
dynamic range 29, 30, 121

edge-blended rear projection 187–8, *188*, 197
edge blending 10, 185–7, *186*
editing software 61, 112
electromagnetic interference (EMI) 137, 154, 179, 183
electromagnetic spectrum 20, *21*
embedded audio 118–19
EMI (electromagnetic interference) 137, 154, 179, 183
emotions 42, 44, 46, 47, 53, 73, 98, 102, 104

emotive design function 95–6
Entertainment Services and Technology Association (ESTA) 160–1
equipment 120–49, *128*
ESTA (Entertainment Services and Technology Association) 160–1
ethernet 138, 155–6, 157, 159, 161, 166, 183, 184, 197
extended reality (XR) 98
eyes, anatomy of 27–29, *28*

fatigue 32–3
fiber optics 24, 183–4, 197
field of view 23, 24
file storage 143
flash memory 141–2, 143
flowchart symbols 55–6
focal length 132–3, 134
focal plane 24, 191
focus 79, 82, 180
frame rate 117
frames 62, 114, 115, 116, 164, 166, 189
frequency 20, *21*, 22, 25, 26, 40, 184, 189; critical fusion (CFF) 33
"frog's-eye view" 84
front projection 170, 171, 172, 173, 175, 177, 180
f-stops 132–3, 134

gain 127, 173–4, 176, 192, 194, 197
geometry 134, 146, 170
graphics card 138, 143, 144, 145
grayscale 71, 86–7

hard disk drive (HDD) 143, 144
hardware 118, 142, 143–5, 149, 163
harmony 79–81, *80*
HDD (hard disk drive) 143, 144
HDMI (High-Definition Multimedia Interface) 19, 135–7, 138–9, 140, 142, 149, 182–3
HDTV (high-definition television) 136
High-Definition Multimedia Interface (HDMI) 19, 135–7, 138–9, 140, 142, 149, 182–3
high-definition television (HDTV) 136
horizon 77, 84
hot spotting 171, 174, *188*

icon 67, 68
idea chart 46, 47, 48, 52, 54, 55
I-frame 115
illuminance 38
illumination 17, 33, 148; determination formulas for 127; measurement of 31–2
image magnification (IMAG) 93
imagery 1–2, 14, 19, 35, 60, 85, 89, 99, 104
index 67, 68
information gathering 167–80, *172*
informative design function 89–93
infrared (IR) 20, *21*, 61, 62
initial site survey 168
inspiration 50, 71, 88–9, 100–1
installation 36, 180–91, *181*, *182*, *186*, *188*
intellectual property 106–7
intensity 22, 28, 29, 30, 32, 33, 40, 185, 187–8
interactive design function 99–100
interactive projection 41, 55, 60, 61, 62, 63, 99, 100
interactive video 58–9, 99–100
inter-frame compression 114
interlaced 195
International Commission for Illumination 30–1
International Organization for Standardization (ISO) 17, 127, 128

Internet Protocol (IP) 157, 158, 161, 162, 183, 190; *see also* TCP/IP
intra-frame compression 114
IP (Internet Protocol) 157, 158, 161, 162, 183, 190; *see also* TCP/IP
IR (infrared) 20, *21*, 61, 62
iris 27, *28*, 29
ISO (International Organization for Standardization) 17, 127, 128

Kelvin 20
keystone 180, 181, *182*
kinesthetic vision 79

lanterna magica 11–13, *13*, 14
latency 60, 139, 154, 183, 184, 189; MIDI Show Control (MSC) 156; MIDI timecode (MTC) 166; Open Sound Control (OSC) 162, 163; User Datagram Protocol (UDP) 158
LED (light-emitting diode) 5; walls 9, 39, 97, 121, 122, 124, 125–6, 129, 173
legibility of text 90
lens aperture 133
lens shift 181, 185
lens throw ratio 133–4
lenses 10, 23, 24, 30, 33, 132–3; throw 134, 171, 180, *188*, 191; zoom 132, 133, 134
light: ambient *see* ambient light; dispersion of 24, 25, 40, 128, *128*, 134; physics of 20–6, *21*, *25*; primary colors of 30, 69; scattering of 24, 25, 26, 40, 176; unwanted 9, 130; visible spectrum of 20, *21*, 70
light meter 32, 128, 168, 197
light pollution *see* light, ambient
light source 2, 5, 9, 11, 20–1, 22, 39–40, 126, 128, 129

light-emitting diode (LED) 5; walls 9, 39, 97, 121, 122, 124, 125–6, 129, 173
lighting design 3
lighting designer 3, 9, 20, 25–6, 39, 68–9, 78, 95–6, 126, 178, 190
line 64, 65–6, *65*
linear perspective 77, 84
linear timecode (LTC) 165
long-distance distribution 182–5
longitudinal timecode (LTC) 165
lossless codecs 114
lossy codecs 114
LTC (linear/longitudinal timecode) 165
lumens 22, 31, 32, 37, 38–9, 40, 127, 128, 130, 131, 171, 173–4, 192
luminance 62, 99, 121, 128
luminous flux 31, 127
lux 22, 31–2

magic lantern 11–13, *13*, 14
matching shadows 78
media players 5, 119, 148, 182; digital 16, 141–2, 149
media program 60, 143
media servers 5, 9, 10, 16, 61, 118, 142, 143, 149; processing visuals in 144–5; and show control 152, 154, 160, 161, 163; software for 145–8
meetings 42, 43, 46–7, 48, 51, 52, 64, 168
memory 2, 72, 113, 141–2, 143, 144, 145
metadata 115, 121
method of distribution 4, 120, 182
MIDI (Musical Instrument Digital Interface) 154–6, 165–6; Show Control (MSC) 154–6; timecode (MTC) 153, 165–6
Mie scattering 26

mismatched aspect ratios *193*
monitors 39, 75, 116, 122–4, 139, 145, 148
mood boards 48, 49–51, 62, 64, 88–9
motion pictures 12–13, 14, 33, 51, 88, 96, 141
motion tracking 61
movement 79, 85–7, *86*
MSC (MIDI Show Control) 154–6
MTC (MIDI timecode) 153, 165–6
multiple controllers 161
multiple displays 5, 7, 10, 33, 122, 135, 138, 146, 188
multiple projectors, designing with 185–9, *186*, *188*
Musical Instrument Digital Interface (MIDI) 154–6, 165–6

narrative 42, 45–6, 53
negative space 66, 67, 77–8
network control 157–63
network protocols 158–62
network switches 161, 183
nit 31
nodes 147, 149, 152
non-traditional projection surfaces 100, 197

off-axis projection *181*, 181–2
Open Sound Control (OSC) 162–3
optical illusions 14, 29, 32–3, *71*, 71, 131
OSC (Open Sound Control) 162–3

particle generation 60–1
passive spacing 78
pattern 64, 72–3
Peirce, Charles Sanders 67
Pepper's ghost 97, 98
perception 30, 34, 64, 92–3, 114, 131, 177, 178; spatial 78–9
persistence of vision 14

perspective 79, 84; horizon 77, 84
P-frames 115
phantasmagoria 11, *13*, 96–7
phénakistiscope 14
photopic vision 28
photoreceptors 27–8, 29, 30, 40
physics of light 20–6, *21*, *25*
pigment 26, 69
pixel density 112, 175, 194, 195
pixel shift 194, 195
pixels 15, 90, 91, 112, 121, 124–5, 175, 194–5
playback 119, 138, 142, 145, 146, 148, 149, 152; devices for 36, 37, 115, 141, 189; video 60, 100, 119, 138, 148, 164
pointillism 33, 101
power 117, 178–80
previsualization 112–13
primary colors 30, 69
production meetings 42, 43, 46–7, 48, 51, 52, 64, 168
programming 10, 36, 56, 147, 149, 152, 163
project management, projection design as 195–7
projected backdrops 35, 94, 180
projected effects 63, 97, 98, 154
projected scenery 94
projection: cinematic 12–13; edge-blended rear 187–8, *188*, 197; front 170, 171, 172, 173, 175, 177, 180; how is it accomplished? 15–18; interactive 41, 55, 60, 61, 62, 63, 99, 100; off-axis *181*, 181–2; rear 23, 25, 170–3, *172*, 177, 187–8, *188*, 197; stacked 187–9; video 11, 43; what makes it right? 37–40; when has it been used? 11–14, *12*, *13*; who is involved? 7–11; why do it? 6–7
projection design: as project management 195–7; using

multiple projectors 185–9, *186*, *188*
projection designer, definition of 2
projection materials, safety concerns of 176–7
projection screens 2–3, 109, 134, 173, 174, 175–6
projection surface 173–8; light on 128, *128*, 130; non-traditional 100, 197; space behind 171; *see also* color; gain; texture; viewing angle
projection system 4, 18, 35, 88, 94, 167, 173, 187
projector: how bright should it be? 127, 131–2, 178; lens 132–4, 189
proportion 79, 83
proposal 46–58
prototypes 11, 57
pupil 27, *28*
pupillary reflex 29
purpose of design 42–3

raster image 91–2
Rayleigh scattering 26
realism 84
rear projection 23, 25, 170–3, *172*, 177, 197; edge-blended 187–8, *188*, 197
reflection 23, 24, *25*, 40, 81, 172; diffuse 23, *25*; specular 23, *25*; total internal (TIR) 24
refraction 23–4, *25*, 28, 40
rehearsals 7, 36, 57, 59, 95, 108, 170, 178, 184
resolution 91, 109–13, **110**, 116, 119, 138, 194–5
retina 27, *28*, 29, 33, 40
rhythm 64, 79, 82, 87
rigging 6, 125, 126, 178–80, 190
rods 11, 27–8, 32, 33–4, 40
royalty-free content 107

sACN (streaming Advanced Controller Network) 153, 160–2, 166
safety concerns, of projection materials 176–7
saturation 22, 30, 69
scattering 24, 25, 40, 176; Mie 26; Rayleigh 26
scenery 3, 7, 12, 20, 93, 94, 95, 158
scenic backdrops 93, 94
scenic design 3, 53, 93–5
scenic designer 3, 8–9, 173, 177, 178–9
scotopic vision 28
script analysis 43–6, 53
SDI (serial digital interface) 139–41, 149, 182–3
secondary color 69
semiotics 67–8
serial digital interface (SDI) 139–41, 149, 182–3
set pieces 59, 61, 76, 189
shadow puppetry 11, *12*
shadows 6, 11, 22, 62, 73, 134, 172, *172*, 180; matching of 78
shape 64, 66–7, *67*
show control 6, 16, 59, 60, 119, 146, 150–66, 190; protocols 152–63; systems 150–1, 151–2, 158–9, 160, 163, 166, 190
sightlines 46, 113, 168–9
signal distribution 10, 19, 137, 169, 179, 197
signal flow 171, 190, 191
signs, use of 67–8
simultaneous contrast 71
single projector 5, 7, 36, 116, 171, 185
slide projector 3, 13, 45
Society of Motion Picture and Television Engineers (SMPTE) 22, 127; timecode 153, 164, 165, 166; video standards 139

software: editing 61, 112; for media servers 145–8
solid-state drive (SSD) 144
source 141
space 64, 75–9, *76*; negative 66, 67, 77–8
spatial acuity 33
spatial dynamics 46
spatial perception 78–9
special effects 45, 119, 150, 153, 155; design function 96–8
specialty theatrical fabrics 177
specifications 191–5, *193*
spectral acuity 33–4
specular reflection 23, *25*
SSD (solid-state drive) 144
stacked projection 187–9
standards 127–8, 134; broadcast 140; cable 138–9; definition of 17; for distribution 135; frame-rate 117; power 117; SMPTE video 139; television 31, 109; theatrical 176–7
stereoscopic vision 78–9
stock content 107–8
storyboarding 48, 51–3, 54, 62, 64, 88–9, 112, 189
storytelling 3, 43, 46, 54
streaming Advanced Controller Network (sACN) 153, 160–2, 166
structural challenges 169–70
Sunday on La Grande Jatte, A 101
symbolism 67, 78, 96
symmetry 81

tag clouds 104–5
TCP/IP (Transmission Control Protocol/Internet Protocol) 153, 158, 161, 163
television 5, 19, 39, 45, 46, 117, 118, 122–3; high-definition (HDTV) 136; standards for 31, 109
temporal acuity 33

test patterns 187, 191
text: in content 103–5; legibility of 90
textural design function 98–9
texture 3, 23, 49, 64, 73–5, *74*
theatrical fabrics 177–8
theatrical standards 176–7
thematic elements 45–6, 47
themes, of productions 11, 43, 44, 89, 96
throw distance 133
throw lenses 134, 171, 180, *188*, 191
throw ratio 132, 133–4
timecode 11, 55, 140, 153, 163–5, 166; longitudinal (LTC) 165; MIDI (MTC) 153, 165–6; SMPTE 153, 164, 165
TIR (total internal reflection) 24
tools 2, 5, 60, 61, 64, 77, 148, 151, 168, 197
total internal reflection (TIR) 24
Transmission Control Protocol (TCP) 158
Transmission Control Protocol/Internet Protocol (TCP/IP) 153, 158, 161, 163

UDP (User Datagram Protocol) 158, 159, 161, 162–3
unwanted light 9, 130
User Datagram Protocol (UDP) 158, 159, 161, 162–3

vanishing point 84
variety 79, 81
vector images 91–2
VESA (Video Equipment Standards Association) 111, 121, 136, 137
VGA (Video Graphics Array) **110**, 136
video black 15, 39, 129, 154, 185
video content 59, 108, 118–19, 136, 157

video data 115, 136
video displays 17, 98, 179
Video Equipment Standards Association (VESA) 111, 121, 136, 137
video extenders 183
Video Graphics Array (VGA) **110**, 136
video playback 60, 100, 119, 138, 148, 164
video projection 11, 43
video projectors 2, 4, 5, 18, 94, 121, 126–32, *128*
video signal 16, 122, 124, 137–8, 140, 145, 183, 188–9
video team 9–11
video transmission 16, 139, 140, 184
video wall 5, 109, 116, 121, 122–4, 126
viewing angle 171, 173, 174, 176, 197
viewing cone *see* viewing angle

virtual reality (VR) 98
visible spectrum, of light 20, *21*, 70
vision: of designer/director 7–8, 10, 42, 46–7, 59, 62, 196; kinesthetic 79; persistence of 14; photopic 28; scotopic 28; stereoscopic 78–9
vocabulary 16–18
VR (virtual reality) 98

wavelength 20, *21*, 25, 26, 28, 31, 33–4, 40, 69
white light 20, 24, 25, 69
wired communication 157, 159
wireless transmission 41, 157, 184
word clouds 104–5
workflow 41, 146, 147

XR (extended reality) 98

zoetrope 14, *15*
zoom lens 132, 133, 134

For Product Safety Concerns and Information please contact our EU representative GPSR@taylorandfrancis.com
Taylor & Francis Verlag GmbH, Kaufingerstraße 24, 80331 München, Germany

www.ingramcontent.com/pod-product-compliance
Lightning Source LLC
Chambersburg PA
CBHW050523170426
43201CB00013B/2067